William Healey Dall, George W. Tryon

American Marine Conchology

Descriptions of the shells of the Atlantic coast of the United States from Maine to

Florida

William Healey Dall, George W. Tryon

American Marine Conchology
Descriptions of the shells of the Atlantic coast of the United States from Maine to Florida

ISBN/EAN: 9783337391416

Printed in Europe, USA, Canada, Australia, Japan

Cover: Foto ©berggeist007 / pixelio.de

More available books at **www.hansebooks.com**

AMERICAN

MARINE CONCHOLOGY:

OR,

DESCRIPTIONS OF THE SHELLS

OF THE

ATLANTIC COAST OF THE UNITED STATES

FROM MAINE TO FLORIDA.

BY

GEORGE W. TRYON, Jr.,

MEMBER OF THE ACADEMY OF NATURAL SCIENCES OF PHILADELPHIA.

PHILADELPHIA:
PUBLISHED BY THE AUTHOR
No. 19 North Sixth Street.

TO THE

MEMORY

OF

THOMAS SAY, AUGUSTUS A. GOULD,

AND

WILLIAM STIMPSON.

"The Almighty Maker has throughout
 Discriminated each from each, by strokes
 And touches of his hand with so much art
 Diversified, that two were never found
 Twins at all points."

COWPER.

"To ask or search I blame thee not; for Nature
 Is as the book of God before thee set,
 Wherein to read his wondrous works.
 But what created mind can comprehend
 Their number, or the wisdom infinite
 That brought them forth, but hid their causes deep."

MILTON.

"There is a pleasure in the pathless woods,
 There is a rapture in the lonely shore,
 There is society, where none intrudes,
 By the deep Sea, and music in its roar;
 I love not Man the less, but Nature more."

BYRON.

PREFACE.

In preparing the present work, it was my purpose to furnish to Conchological students and to sea-shore collectors succinct and plain descriptions, illustrated by characteristic figures of the American Marine Mollusks inhabiting our coast from Maine to Florida. I could not undertake, within the limits of a single volume, to give a complete portraiture of each species, or to present its entire bibliography, and I have therefore cultivated brevity—I hope in no case at the expense of lucidity.

Following the plan adopted in my previous memoirs on American Conchology I have prepared copious analytical tables of families, genera, and species, presenting their prominent distinctive characters at a glance, and thus greatly facilitating their correct determination. In my classification I have not always followed the most approved modern systematists, simply because it appeared to me to be very unnecessary in a work of such partial character as the present one to introduce a host of systematic divisions, where an older, more simple, and more generally comprehensible method would subserve my purpose.

Space could not be found for full notes of the habits of all of our species heretofore observed, but I think that I have described a sufficient number of them to give the reader a reasonably good idea of their appearance, mode of life, etc., and to incite collectors to observe and to study the living specimens. The limits of bathymetrical as well as of geographical distribution of the species have been carefully stated, and collectors may expect their occurrence in suitable situations at all intermediate localities and depths.

With regard to the geographical limits which I have assigned to my work, I would explain that, northward of the Canadian waters, the coast has not been sufficiently explored to lead to the conjecture that we have been made acquainted with the moiety of its species; moreover it is very unlikely that collectors will be supplied with many of these Arctic species. Of those known to science a considerable portion extend southwards to Cape Cod, and are consequently herein described. Our southern limit, on the other hand, will include a considerable number of tropical forms which, dispersed and protected by the warmth of the Gulf Stream, have spread to various portions of our southern coast. I could not extend my limits southward to the southern coast of Florida or that of the Gulf of Mexico without including the great West Indian province, which would have enlarged my work to five or six times its present bulk. This portion of our marine molluscous fauna, like that of our Pacific coast, may be advantageously presented in separate treatises.

It gives me great pleasure to acknowledge my obligations to many friends for their active interest in my work; I am particularly indebted to Mr. JOHN H. THOMSON, of New Bedford, Mass., and Dr. H. C. YARROW, of Fort Macon, N. C., for extensive collections from localities rich in species. I have dedicated my work to the memory of the three deeply lamented naturalists to whom we are principally indebted for our knowledge of American Marine Mollusca. Two of them, Dr. GOULD and Dr. STIMPSON, were personal friends, and from the time when they fostered my first efforts in natural science almost to the period of their deaths I never applied to them in vain for assistance and advice in my Conchological studies.

The generic and specific descriptions used in this book are principally copied from "Woodward's Manual of the Mollusca," Adams' "Genera of Recent Mollusca," De Kay's "Mollusca of New York," Gould's "Invertebrata of Massachusetts," and from the original diagnoses of Say, Adams, Mighels, Stimpson, and others. I think that it is unnecessary to make any apology for using so extensively the language of other writers in preference to preparing original descriptions—the latter course I have adopted whenever I believed that I could improve or correct the original description. I trust that this general acknowledgment will save me from the reproach of egotism in avoiding the use of cumbersome and unsightly quotation marks.

Upon the completion of each of my former volumes I have promised myself that it should be my last book; that pledge I am again impelled to violate. Those alone, who labor in a similar field, can understand how imperatively science requires her votaries to publish her truths for the instruction of the world.

GEO. W. TRYON, JR.

JANUARY 1, 1873.

MARINE MOLLUSCA OF THE UNITED STATES.

Class **CEPHALOPODA.**

HEAD large, separate from the body; eyes large, complex, lateral: ears developed; mouth armed with two horny or shelly jaws edged with fleshy lips, surrounded by eight or ten fleshy arms and furnished with an entire or slit tube or siphuncle. Body ovate, roundish or cylindrical, open in front, containing the viscera and one or two pairs of internal symmetrical gills; naked; surrounded by a thin shell with a single cavity; or partly or entirely contained in the last chamber of a chambered shell furnished with a siphon passing from chamber to chamber. Unisexual. Animal free, walking head downwards by means of its arms, or swimming in the sea, propelled by the water from the siphon tube.

ORDER OCTOPODA.—Body naked; head separate, with *eight* fleshy arms, furnished with *sessile* cups or suckers, *without* any horny rings; eyes *fixed* in the skin. Siphuncle entire; gills two. *No internal or external shell.*

ORDER DECAPODA.—Body naked; head separate, with *ten* fleshy arms, two of them longer, furnished with *pedunculated* cups or suckers, *with* horny rings; eyes *free* in their orbits. Siphuncle entire; gills two. *Having an internal shell or pen.*

ORDER **OCTOPODA.**

Family OCTOPODIDÆ.

Arms subulate. Mantle supported by fleshy bands. No cephalic aquiferous apertures.

In the genera comprising this family the arms are similar, elongated, and united at the base by a web. The animal is chiefly littoral, inhabiting the temperate and tropical zones. It escapes capture by varying its tints according to the nature of the ground over which it passes, and eludes its enemies, when pursued, by discoloring the water around it with the contents of its ink-bag.

Synopsis of Genera.

Arms with two rows of cups. Body round, without fins.   OCTOPUS.
Arms with one row of cups, bearded and united by a broad web. Body finned.   CIRRHOTEUTHIS.

Genus OCTOPUS, Cuvier.
Regne Animal, ii. 1817.

1. O. RUGOSUS, Bosc. Fig. 1.
Act. de la Soc. d'Hist. Nat., t. 15, f. 1, 2. 1792.

Sepia granulosa, Bosc. Buff., Vers. i. 47. 1802.
Sepia granulatus, Lamarck, Mem. Soc. Hist. Nat., i. p. 2. 1799.
Octopus Barkeri, Ferussac, Orb. Sal. Céph. 54, No. 3. 1826.
Octopus Americanus, Blainville, Dict. Sc. Nat. xliii. 185. 1826.

Body oval, purse-shaped, large, with a deep groove above. Head, arms, and upper part of the body covered with roundish tubercles. Head short, warty; ocular beard, one, elongated. Arms short, thick, conical, their proportional lengths 4, 3, 2, 1. Cups large, of upper part of arm rather smaller, lowest one-rowed. Web short. When alive, violet-brown, white beneath; side of the arms netted with brown lines.

North Carolina to South America; also Indian Ocean. ?

Genus CIRRHOTEUTHIS, Eschricht.
Nov. Act. Nat. Cur., xviii. 625. 1836-8.

1. C. MULLERI, Eschricht. Fig. 2.
Nov. Act. Nat. Cur., xviii. 1836-8.

Body smooth, oblong, three-lobed. Fins longer than broad, blunt, depressed. Eyes very small. Arms equal, quadrangular. Cups very small, oval, about thirty; beards between the cups filiform.

Arctic America.

ORDER DECAPODA.

* Shell internal, solid, horny.

Family LOLIGOPSIDÆ. Eyes naked. Mantle supported by two internal fleshy bands. Siphuncle simple. Fins semicircular.
Family ONYCHOTEUTHIDÆ. Eyes naked, with a sinus above.

Mantle furnished with three internal cartilages, one dorsal and two ventral. Siphuncle with a valve. Fins angular.

Family LOLIGINIDÆ. Eyes covered with skin, simple. Mantle with three internal cartilages, one dorsal and two ventral.

 * Shell internal, calcareous, spiral, chambered, siphunculated.

Family SPIRULIDÆ. Eyes covered with the skin, with a lower eyelid. Mantle free all around.

Family LOLIGOPSIDÆ.

The members of this family have the eyes pedunculated and not covered by a skin; the fins are caudal, terminal, and semi-circular; the body is membranaceous, semipellucid, elongate, and tapering behind. They inhabit the high seas, and are powerful swimmers.

Genus LOLIGOPSIS, Lamarck.
Extr. de Cours. 1812.

Leachia, Lesueur, Journ. Philad. Acad., ii. 89. 1821.

Body very long, conical, attenuated. Shell flexible, slender, keeled above; very narrow, lanceolate, thickened at the tip.

1. L. PAVO, Lesueur. Fig. 3.
 (*Loligo.*) Journ. Philad. Acad., ii. 96. 1821.

Body smooth, spotted with red. Fins terminal, short, soft, narrow, outline together heart-shaped, not notched in front. Sessile arms slender, short, three upper pairs rounded. Cups much depressed, broad, oblique; rings smooth exteriouly, inner edge divided into square teeth. Tentacular arms slender. Shell elongate, very thin, nearly gelatinous, attenuated anteriorly, lanceolate posteriorly.

Arctic Seas, New England, New York, Madeira.

Family ONYCHOTEUTHIDÆ.

Synopsis of Genera.

 * *Tentacular Arms with Hooks. Sessile Arms with Cups and Rings.*

Club of tentacular arm with hooks. Shell lanceolate, pennate, sides thin.

        ONYCHOTEUTHIS.

Club of tentacular arm with hooks on the centre, and with two rows of small cups on each side. Shell lanceolate, pennate. ONYCHIA.

 * *Tentacular and sessile arms with cups and horny rings.*

Fins rhombic, posterior, caudal. Internal cartilage of mantle dilated below. Shell narrow, dilated in front with one central and two marginal ribs.

OMMASTREPHES.

Genus ONYCHOTEUTHIS, Lichtenstein.

Berl. Acad. 1818.

Body elongate, subcylindrical, smooth, acuminated behind. Mantle with an elongated, narrow, prominent, longitudinal ridge, fitting into a similar groove on the base of the siphuncle. Head large, rather depressed, with three to eleven longitudinal ridges, and edged behind by a transverse ridge. Eyes large, lateral; sessile arms angular; third or fourth pair with a keel or fin; cups in two alternating lines; rings convex and toothless. Tentacular arms partly retractile, strong; club enlarged, with a rounded group of small sessile cups at each end, and two series of claw-like hooks, the outer series largest. Siphuncle very short, lodged in a cavity, with two superior muscular bands.

1. O. BANKSII, Leach. Fig. 4.

(*Loligo.*) Zool. Miscel., iii. 141, sp. 4. 1817.

Onychoteuthis Bergii, Lichtenstein, Zool. Mus. Berlin, 1592, t. 19, f. a. 1818.
Sepia loligo, Fabricius, Faun. Grœnl. 359.
Onychoteuthis Fabricii, Lichtenstein. Isis. t. 19. 1818.
Onyka angulata, Lesueur, Journ. Philad. Acad., ii. 99, t. 9, f. 3, 184 ; ii. 296, fig. 1822.
Loligo felina, Blainv. Dict. Sc. Nat.; xxvii. 139. 1823.
Onychoteuthis Molinœ, Leach, Berl. Trans., t. 4. 1818.
Loligo uncinatus, Quoy & Gaimard, Zool. Uranie, i. 410, t. 66, f. 7. 1838.
Onychoteuthis Lessonii, Fer. Orb. Ceph. 61, No. 6. 1825.
Onychoteuthis Fleurii, Renaud. Lesson, Centurie Zool. 61, t. 17.

Head, with eleven longitudinal, small, prominent ridges. Fins rhomboidal. Sessile arms unequal, in length, 2, 3, 4, 1. Cups pear-shaped. Tentacular arms very extensile, with double series of hooks; the basal group consisting of seven or eight open, and the same number of closed cups. The special group of sixteen or seventeen cups all open; hooks *twenty to twenty-two*, in two rows, those of the outer row largest.

Arctic Seas to West Indies, Africa, Indian Ocean.

2. O. BARTLINGII, Lesueur. Fig. 5.

(*Loligo.*) Journ. Philad. Acad., ii. 95, fig. 1821.

Onychoteuthis Lesueurii, Ferussac, Céph. Acét., t. 4.

Body elongate; back with a central, transparent line over the keel of the shell. Fins rhombic. Sessile arms slender; dorsal pair rounded externally, with a slight fin on the upper part; the second, third, and fourth pairs finned on the outer side nearly the whole length. Tentacular arms with *six* large hooks. Shell dark-brown, lanceolate, pennate, with a short central keel above and ridge beneath, thin.

Gulf Stream, Indian Ocean, New Zealand.

Genus ONYCHIA, Lesueur.

Journ. Philad. Acad., i. 98, 1821 ; ii. 296, 1822.

In this genus the body is red and spotted; the tentacular arms scarcely enlarged at the ends. Like most other genera of this family, and other pelagic forms, it is crepuscular, darting along the surface of the ocean toward nightfall, and preying upon small fishes, floating crustacea and acalephæ that swim near the surface.

1. O. CARDIOPTERA, Péron.　Fig. 6.

(*Loligo.*) Voy. Atlas, t. 60, f. 5.　1804.

Sepiola cardioptera, Lesueur, Journ. Philad. Acad., ii. 100.　1821.
Onychia caribæa, Lesueur, ibid., ii. 98, t. 9, f. 1, 2.　1821.
Onychoteuthis Leachii, Férussac, Céph. Act. Onych., t. 5, f. 4, 7.

Body large, oblong, narrowed and prolonged behind. Sessile arms unequal, relative lengths 3, 2, 4, 1; cups in two alternating lines. Shell rather broad, sides rounded.

Southern Atlantic Ocean ; Gulf of Mexico.

Genus OMMASTREPHES, Orbigny.

Moll. Viv. et Fos., i. 412.　1845.

In this genus the sessile arms are conical, subulate, upper and lower quadrangular, the others triangular, compressed, unequal; relative lengths 3, 2, 1, 4; the cups are very oblique, fleshy, distinct; the rings oblique and toothed. Tentacular arms not retractile, short, strong, thick, with a slight longitudinal ridge, scarcely enlarged at the end, webbed at the tip.

Living in the high seas in large troops, nocturnal; the food of cetacea and pelagic birds. The sailors call them "sea-arrows" or "flying-squids," from their habit of leaping out of the water, often to such a height as to fall on the decks of vessels.

1. O. SAGITTATUS, Lamarck.   Fig. 7.

(*Loligo.*)   Mém. Soc. Hist. Nat., Paris, xiii. 1799.

Loligo illecebrosa, Lesueur, Journ. Philad. Acad., ii. 95, plate. 1821.
Loligo harpago, Ferussac, Dict. Class. Hist. Nat., iii. 67. 1823.
Loligo Brongniartii, Blainv., Dict. Sc. Nat., xxvii. 142. 1823.
Loligo piscatorum, La Pylaie, Ann. Sc. Nat., iv. 319. 1825.
Loligo Coindetii, Verany, Mem. Acad. Sc. Torino, t. 1, f. 4. 1837.

Head large. Body elongate, cylindrical, opaque, fleshy, smooth
above and below. Tentacular arms with *eight* rows of numerous
small cups at the extremities. Shell narrow, elongate; lateral
ribs largest; apical cone large.

This beautiful animal is occasionally seen on all parts of the
shore of Massachusetts. But it is especially abundant about
sandy shores, as at Cape Cod. At Provincetown I have seen them
stranded upon the beach at low tide, in great multitudes. Their
usual mode of swimming is by dilating their sack-shaped body
and filling it with water. The body is then suddenly contracted
and the water forcibly ejected, so as to propel them backwards
with great rapidity. So swift and straight is their progress that
they look like arrows shooting through the water. Whenever
they strike the shore they commence pumping the water with increased violence, while every effort only tends to throw them still
further upon the sand, until they are left high and dry. The body
is beautifully spotted with colors, which seem to vary with the
emotions of the animal. At one moment they are a vivid red, at
the next a deep blue, violet, brown, or orange. They devour immense numbers of small fish, and it is amusing to watch their
movements and see how, at a distance of several feet, they will
poise themselves, and in an instant, with the rapidity of lightning,
the prey is seized in their long arms and instantaneously swal-
lowed. They, in their turn, are devoured by the larger fishes, and
are extensively used for bait in the cod-fishery. (*Gould, Invert.
Mass.*)

     Atlantic Ocean from Newfoundland southwards; Mediterranean.

2. O. BARTRAMII, Lesueur.   Fig. 8.

(*Loligo.*)   Journ. Philad. Acad., ii. 90, t. 7. 1821.

Loligo sagittatus, Blainv., Dict. Sc. Nat., xxvii. 140. 1823.
Loligo vitreus, Rang, Mag. Zool. 71, t. 86. 1837.
Ommastrephes cylindricus, Orb. Voy. Am. Merid. 54, t. 3, f. 3, 4.

Head short. Body elongate, cylindrical, acuminate posteriorly, truncated anteriorly, longitudinally adorned above with a violet zone. Tentacular arms with *two* series of small cups at the extremities. The second and third pairs of sessile arms with a broad membranaceous fin on the inner edge of the ventral side. Shell thin, elongated.

Atlantic Ocean, Mediterranean, Cape of Good Hope.

Family LOLIGINIDÆ.

The eyes are covered by the skin, without lids. Sessile arms with two rows of cups. Rings not convex externally, provided with a narrow, prominent edge on the middle of their width. Tentacular arms only partly contractile.

Synopsis of Genera.

Head separate from the body. Fin terminal, rhombic.

Cups of sessile arms in two rows; lateral membranes with cups on the angles. Shell as long as the back. LOLIGO.

Head attached to back of mantle by a band. Fins short, dorsal, not terminal. Shell narrow, with a central and two marginal ribs. SEPIOLA.

Genus LOLIGO, Lamarck.

Mém. Soc. Hist. Nat. 1799.

These animals pursue their prey on the bosom of the ocean, swimming with great rapidity; fish and pelagic crustaceans, ianthinæ and other oceanic mollusca, constitute their food; many individuals frequently unite and hunt in companies; their favorite time for scouring the surface being the evening after sunset. It is the favorite food of the cod, and with it one-half of all the cod taken at Newfoundland is caught. When the European species, *L. magna*, approaches the coast in vast shoals, five hundred sail of English and French ships engage in the fishery for bait. During violent gales hundreds of tons of them are often thrown up together in beds on the English coast. There are numerous species, inhabiting all seas.

1. L. BREVIS, Blainville. Fig. 9.
   Journ. de Phys. 1823.
Loligo brevipinna, Lesueur, Journ. Philad. Acad., iii. 282, t. 10, f. 1–3.
   1824.

Body cylindrical, obtuse posteriorly. Fins short, transversely oval. Shell dilated, very broad, central keeled, narrow in front.

Delaware Bay (Lesueur), *Fort Macon, North Carolina; Southern Atlantic Ocean to Brazil.*

2. L. PUNCTATA, De Kay. Figs. 10, 11.

Moll., New York, 3, t. 1, f. 1. 1843.

Body cylindrical, thick, somewhat tapering posteriorly. Fins broadly rhomboidal, nearly half the length of the body. Cups irregularly placed, numerous.

This beautiful squid is nearly allied to the *L. Pealii* of Lesueur, but this latter has the suckers arranged in two regular series. It has also a membrane along the lateral edges of the arms, and an acute termination of the caudal extremity.

Fig. 11 represents a bunch of egg-cases and an embryo, highly magnified. (Copied from De Kay.)

New York; Connecticut.

3. L. PEALII, Lesueur. Fig. 12.

Journ. Philad. Acad., ii. 92, t. 8, f. 1, 2. 1821.

Body elongate, subconical, acuminated posteriorly. Fins rhomboidal, thick, occupying about three-fifths of the length. Sessile arms long; cups very oblique, in two rows. Shell lanceolate, narrow.

Entire Atlantic Coast.

Genus SEPIOLA, Rondelet.

Piscis et Aquat., i. 510. 1554.

The body in this genus is round and purse-like, and the short dorsal fins are rounded and contracted at the base. Distribution universal.

1. S. ATLANTICA, Orb. et Ferussac. Fig. 13.

Céphal. Acet., 235, t. 4, f. 1-12. 1839.

Sepiola vulgaris, Gervais & Van Beneden, Bull. Acad. Bruxelles, iv. No. 7. 1849.

Sepiola oceanica, Orb. Moll. Viv. et Foss, t. 10, f. 13. 1845.

Fins oval, far apart. Sessile arms short, unequal, with two rows of cups; the lower pair with eight rows of smaller cups at the tip; proportionate lengths 3, 2, 4, 1; cups small, oblique, in two series; lateral arms larger. Tentacular arms long. Shell linear, narrow, gradually enlarged upwards and spathulate behind the tip; sides thickened.

Atlantic Ocean.

Family SPIRULIDÆ.

This family differs from those preceding it in having a calcareous spirally coiled and concamerated shell. There is but one recent genus, *Spirula*, and only one species.

Genus SPIRULA, Lamarck.

Extr. de Cours. 1799.

1. S. PERONII, Lamarck. Figs. 14, 15.
   Anim. sans Vert., vii. 601. 1822.

Spirula fragilis, Stimpson, Check List, 6. 1860.

Shell white, nacreous, coiled in two or three turns, which do not touch each other, something like a ram's horn. The surface exhibits constrictions, at short intervals, each of which corresponds to an internal partition, so that the whole shell is divided off into chambers, having a tube at one side, so that the whole are in communication. Inhabits the open sea all over the world, and is cast upon the shores during storms.

CLASS GASTEROPODA.

Head distinct, furnished with eyes and tentacles. Body usually protected by a conical or spiral shell. Lower part of body formed into a thickened, expanded, creeping disk or foot.

ORDER I. PROSOBRANCHIATA.

Animal creeping or swimming, protected by a shell usually large enough to contain it. Branchiæ plume-like, situated in advance of the heart. Sexes distinct.

A. Siphonostomata. Animal provided with a siphon and having a canaliculated shell. Carnivorous. The shell is spiral, and its axis is usually imperforate. Operculum lamellar, horny.

B. Holostomata. Respiratory siphon wanting, or represented by a mere lobe in the collar of the mantle. Shell spiral or limpet-shaped, generally somewhat globular, and with an entire rounded aperture. The gills are plume-like, placed obliquely across the back, or attached to the right side of the neck. These mollusks inhabit the sea or fresh water, and all of the latter, as well as a portion of the former, are phytophagous.

ORDER II. **PULMONIFERA**.

Lung breathers.   Land and fresh-water snails.

ORDER III. **OPISTHOBRANCHIATA**.

Shell rudimentary or wanting. Branchiæ arborescent, not protected, but more or less completely exposed, on the back or sides of the body near the rear end. Sexes united.

A few species, *Bulla* for example, have a shell enveloped in the mantle.

ORDER IV. **PTEROPODA**.

Marine animals swimming by the aid of a pair of wing-like fins proceeding from the sides of the mouth or neck. Shell glossy and translucent. Sexes united. Pelagic.

ORDER I. **PROSOBRANCHIATA**.

Section A. SIPHONOSTOMATA.

Carnivorous *Gasteropoda*. Shell spiral, axis usually imperforate; aperture notched or produced into a canal in front. Operculum horny, lamellar.

* *Lip of aperture expanded.*

Family STROMBIDÆ. Shell with an expanded lip, deeply notched near the canal. Operculum claw-shaped, serrated on the outer edge.

** *Lip of aperture not expanded.*

Family MURICIDÆ. Shell with a straight anterior canal; aperture entire behind.

Family BUCCINIDÆ. Shell simply notched in front; or with a short canal abruptly reflected, producing a kind of varix on the front of the shell.

Family CONIDÆ. Shell inversely conical or subfusiform; aperture long and narrow; outer lip notched at or near the suture. Columella without plaits.

Family VOLUTIDÆ. Shell porcellanous, turreted or convolute; aperture notched in front; columella obliquely plaited.

Family CYPRÆIDÆ. Shell porcellanous, convolute; spire concealed; aperture as long as the shell, narrow, channelled at each end; outer lip of adult thickened, inflected.

Family STROMBIDÆ.

Animal furnished with large eyes, placed on thick pedicels; tentacles slender, rising from the middle of the eye-pedicels. Foot narrow, ill-adapted for creeping.

The strombs are carrion-feeders, and for molluscous animals, very active; they progress by a sort of leaping movement, turning their heavy shells from side to side. Their eyes are more perfect than those of the other gasteropods, or of many fishes.

Genus STROMBUS, Linn.

Syst. Nat., edit. ii. 64. 1740.

Shell rather ventricose, tubercular or spiny; spire short; aperture long, with a short canal above and truncated below; outer lip expanded, lobed above, and sinuated near the notch of the anterior canal.

The fountain-shell of the West Indies, *S. gigas*, L., is one of the largest living gasteropods, weighing sometimes four or five pounds. Immense quantities of it are annually used for the manufacture of cameos, and for the porcelain works; 300,000 were brought to Liverpool from the Bahamas in one year. There are about seventy-five species; inhabiting all tropical seas.

1. S. ALATUS, Gmelin. Fig. 16.

Syst. Nat. 3513. 1790.

Strombus pyrulatus, Lamarck, Anim. s. Vert. vii. 205. 1822.

Shell ovately conical, rather stout, spire acuminated, whorls smooth, conspicuously grooved at the base and towards the apex, concave round the upper part, noduled at the angle, nodules rather small, columella very callous, lip winged, interior of the aperture wrinkled towards the lip; chestnut-brown, columella, and interior of the aperture deep blackish-chestnut, sometimes carnelian-red.

Beaufort, North Carolina, to West Indies.

Genus APORRHAIS, Dillwyn.*

Philos. Trans., ii. 395. 1823.

Shell with an elongated spire; whorls numerous, tuberculated; aperture narrow, with a short canal in front; outer lip of the adult expanded and lobed or digitated; operculum pointed, lamellar.

---

* On the authorship of this name see W. M. Gabb in Am. Journ. Conch., iv. 143. 1868.

2

Animal with a broad short muzzle; tentacles cylindrical, bearing the eyes on prominences near their bases, outside; foot short, angular in front; branchial plume single, long.

Subgenus Arrhoges, Gabb.

Am. Journ. Conch., iv. 145. 1868.

Shell fusiform, anterior canal nearly obsolete, no posterior canal, outer lip expanded, simple.

1. A. OCCIDENTALIS, Beck.   Fig. 125.

Guérin, Mag. Zool., t. 72. 1836.

Whorls eight to nine, convex, with numerous waving vertical folds and regular conspicuous revolving lines; lip expanded, with a blunt process above.   Epidermis thick and dusky; beneath bluish-white.   Length 2.25; width 1.5 inches.

Maine, northwards.

Family MURICIDÆ.

Animal with a broad foot; eyes sessile on the tentacles, or at their base; branchial plumes two. Lingual ribbon long, linear. Teeth in three series (1·1·1) the central broad, the lateral versatile. Mantle inclosed, the margin producing varices at intervals across the shells, and extended in front, forming a straight, more or less elongated siphon.   Foot simple in front.

Synopsis of Genera.

1. *Not umbilicated.   Outer lip thickened.*

Lip without folds ; whorls with varices.

Shell ornamented with *three or more* continuous longitudinal varices; aperture rounded; beak often very long; canal partly closed; operculum concentric; nucleus sub-apical.   MUREX, Linn.

Shell with *two* rows of continuous varices, one on each side.

RANELLA, Lam.

2. *Not umbilicated.   Columellar lip with folds ; no varices.*

Shell fusiform, elongated; not cancellated; whorls round or angular; canal open; columellar lip tortuous with several oblique folds. Operculum claw-shaped.   FASCIOLARIA, Lam.

Shell cancellated; aperture channelled in front; columella with several strong oblique folds; no operculum.   CANCELLARIA, Lam.

3. *Umbilicated.   Shell spirally furrowed ; columella without folds.*

Thin, *umbilicated ;* spirally furrowed; the ridges with epidermal fringes; columella obliquely truncated; operculum lamellar, nucleus external.

TRICHOTROPIS, Brod.

4. *Not umbilicated. Shell pear-shaped ; fusiform ; outer lip thin ; columella smooth.*

Pear-shaped ; spire short ; whorls angular above ; canal long, curved, open ; operculum with apical nucleus.   Sycotypus, Browne.

Pyriform, light, ventricose, ribbed, and cancellated ; spire very short ; aperture large ; columella simple ; canal straight, elongated ; outer lip thin, entire.   Ficus, Klein.

Fusiform ; spire conical ; canal straight, long ; operculum ovate, curved ; nucleus apical.   Fusus, Lamarck.

Genus MUREX, Linn.

Syst. Nat., edit. x., 746. 1758.

The murices appear to form only one-third of a whorl annually, ending in a varix, but some species form intermediate varices of less extent.   There are over two hundred species, of world-wide distribution.

1. M. spinicostata, Valenciennes. *1843,* Fig. 17. *not of Brown, 1827.*
    Kiener, Icon. Coq., 49, t. 41, f. 1. — *M. fulvescens Sby. 1840,*

Shell pear-shaped, sometimes oblong, sometimes shorter and ventricose; whorls biangulated at the upper part, transversely ridged and striated ; ridges minute, narrow, rather superficial ; six-or-seven varicose veins armed with frond-like spines throughout ; spines canaliculated, slightly curved, the basal, and those upon the posterior angles, longer, sharp-pointed ; white, ridges reddish-brown, stained here and there with small crescent-shaped spots of a deeper color; interior of the aperture white, stained in places with ruddy spots; canal rather elongated; sometimes shorter, recurved.

Beaufort, North Carolina, to Gulf of Mexico.

The North Carolina specimens are generally worn, evidently dead shells. This is probably the most northern limit of the species.

Subgenus Urosalpinx, Stimpson.

Am. Journ. Conch., i. 58. 1865.

Shell elongated, oval or short fusiform, longitudinally ribbed or undulated and spirally striated: aperture with a short canal. Operculum semi-cordate, with the nucleus at the outer edge a little below the middle.

1. U. cinereus, Say.   Fig. 42.
    Fusus. Journ. Phil. Acad., ii. 236. 1821.
Buccinum plicosum, Menke, Synopsis, 2d edit. 59. 1830.

Shell oval, fusiform, with five or six convex volutions crossed by a dozen rib-like undulations, and with numerous revolving lines; mouth oval, brownish within, canal short. Low water to fifteen fathoms.

Length 25; diam. 15 mill.

Inhabits the whole coast.

Genus RANELLA, Lam.
Extr. d. Cours. 1812.

Shell ovate or oblong, compressed, with two rows of continuous varices, one on each side; aperture oval; columella arcuated; canal short, recurved; outer lip crenulated.

The species are mostly tropical. They crawl with considerable animation and rapidity.

1. R. CAUDATA, Say. Fig. 18.
Journ. Acad. Nat. Sci., Philad., ii. 236. 1822.
Fusus pyruloides, De Kay. Moll., New York, 147, t. 9, f. 191. 1843.
Eupleura caudata, Say. Stimpson, Am. Journ. Conch., i. 58. 1865.

Shell solid, whorls five, with nine stout, vertical ribs, besides the varices, crossed by numerous revolving lines. Lip thick, bordered within by raised granules. Reddish-brown, white or bluish-white within. Operculum chestnut.

Length 1 inch; width .65 inch.

The head and tentacles and the siphon are nearly white, the foot light yellow.

Massachusetts to Georgia.

Genus FASCIOLARIA, Lamarck.
Prodr. 1799.

Shell fusiform; spire acuminated; aperture oval, elongated, as long as the spire; siphonal canal straight; columella smooth, with a few oblique plaits at the forepart; outer lip internally crenate. Operculum claw-shaped.

There are over one hundred species; tropical or subtropical in distribution.

1. F. GIGANTEA, Kiener. Fig. 19.
Icon. Coq., Viv. Fasciolaria, p. 5, t. 10 and 11.

Shell symmetrically fusiform, spire acuminately turreted; whorls somewhat obsoletely obtusely ridged throughout, scarcely angulated around the upper part, armed with large swollen tubercles; fleshy-white, covered with a subtransparent yellowish-chestnut horny epidermis, columella and interior of the aperture reddish.

This is the largest species of gasteropod known, attaining to the length of from one to two feet.

Southern Coast, from North Carolina to Florida.

2. F. TULIPA, Linnæus. Fig. 20.

(*Murex.*) Syst. Nat., edit. xi. 754. 1758.

Fusiform, ventricose, spirally, irregularly grooved, suture crenated. Bluish-white, variegated with chestnut or olive blotches, sometimes spirally encircled with black lines; flesh color within.

North Carolina (rare) to West Indies.

3. F. DISTANS, Lamarck. Fig. 21.

Anim. sans Vert. vii. 119. 1822.

Fusiform, ventricose, smooth, polished, spirally ridged at the base; whitish, variegated with chestnut blotches, and encircled with distant black lines.

Distinguished from *F. tulipa* by its smaller size and more distant revolving black bands. It is probably not distinct.

North Carolina to West Indies.

4. F. LIGATA, Mighels and Adams. Fig. 22.

Bost. Journ. N. Hist., iv. 51, t. 4, f. 17. 1842.

Ptychatractus ligatus, Stimpson, Am. Journ. Conch., i. 590. 1865.

Shell elongate, fusiform, thick, reddish; whorls six, convex, with revolving ribs; spire acuminate, suture strongly impressed; aperture ovate-elongate; lip crenate; columella with two folds.

Gulf of St. Lawrence.

This is certainly an aberrant form, and its pertinence to the genus is extremely doubtful.

Genus CANCELLARIA, Lamarck.

Prodrom. 1799.

According to M. Deshayes, the Cancellaria is a vegetable feeder. The typical species have strong oblique plaits on the columella; they range from low water to forty fathoms and inhabit warm latitudes. About seventy-five species have been described.

1. C. RETICULATA, Linnæus. Fig. 23.

(*Voluta.*) Syst. Nat., edit. xii. 1190. 1767.

Shell oblong-turbinated, imperforated, solid, spire exserted; whorls convex, everywhere obtusely reticulated; whitish, banded and variegated with red-brown; aperture rather narrowly ovate, interior strongly ridged, plaits two, very prominent.

Southern Coast from North Carolina (rare) to West Indies.

Subgenus **Admete,** Kroyer.

Möller, Naturhist. Tidskr., iv. 88. 1842.

Shell ovate, thin, covered with an epidermis; spire acute, last whorl ventricose; aperture oval, chanelled anteriorly; columella with a few obsolete rudimentary folds; outer lip thin, simple, acute.

This group appears to represent *Cancellaria* in the northern seas in the same manner that *Trophon* represents *Murex,* and *Bela* certain species of *Mangelia.*

2. C. VIRIDULA, Fabricius. Fig. 24.

(*Tritonium.*) Fauna Grœnlandica, 402. 1780.

Cancellaria buccinoides, Couthouy, Bost. Journ. Nat. Hist., ii. 105, t. 3, f. 3.
Cancellaria Couthouyi, Jay, Cat. 1839.
Admete viridula, Stimpson, Check Lists, No. 6.

Shell oblong, longitudinally obscurely ribbed, spirally lined, spire acuminated; whorls rounded, suture strong; aperture short, smooth, columella obsoletely plaited; white, under a pale green epidermis. Ten to forty fathoms.

Massachusetts to Arctic Ocean.

Genus **TRICHOTROPSIS,** Brod. & Sowb.

Zool. Journ., iv. 373. 1826.

Shell thin, ventricose, keeled, umbilicated; aperture longer than the spire, compressed into a partial canal in front; epidermis horny, rising into hairs at the angles of the shell; operculum horny, nucleus at one side.

There are about a dozen species, principally of arctic distribution.

1. T. BOREALIS, Sowerby. Fig. 25.

Zool. Journ., iv. 373, t. 9, f. 6, 7. 1826.

T. costellatus, Couthouy, Bost. Journ. Nat. Hist. ii. 108, t. 3, f. 2.

Shell ovate-rhomboidal; whorls four, the last very broad, and encircled by four or five, and the others by two prominent, fringed ribs, and crossed by minute and regular elevated lines. Fifteen to twenty fathoms.

From Massachusetts northward, Northern British Coasts.

Genus **SYCOTYPUS**, Browne.

Nat. Hist., Jamaica, 406. 1756.

Busycon, Bolten, Mörch, Yoldi, Cat. 104. 1852.
Fulgur, Montfort, Conch. Syst., ii. 502. 1810.
Pyrula, Lam. (part.) Prodr. 1799.

Shell pyriform, thin, last whorl large, nodulous or spinose; spire very short; aperture large, subtriangular; canal open, elongated, entire at the forepart; inner lip concave, with a single fold anteriorly; outer lip internally striated.

The species are mostly American and are principally southern. Recent authorities have separated the shells with canaliculated sutures under the generic name *Sycotypus*, leaving the other species to constitute the genus *Fulgur* = *Busycon*. As I fail to appreciate the necessity for this division I have reunited them under the name of *Sycotypus*.

* *Canaliculate around the suture.*

1. S. CANALICULATUS, Linnæus. Figs. 26, 27.
(*Murex.*) Syst. Nat., edit. xii. 1222.

Pyrula canaliculata, Brug., Ency. Meth. Vers. iii. 866, t. 436, f. 3.
Pyrula spirata, Kiener, Iconog., t. 10, f. 1.
Sycotypus canaliculatus, Gill., Am. Journ. Conch. iii. 149. 1867.

Shell large, pear-shaped, covered with revolving lines, and a hispid epidermis; lower whorl tumid, ending in a long canal; a nodular keel crowns the flattened summit of each whorl, and there is a deep and broad channel at the suture.

Fig. 27 represents a string of egg cases.

Cape Cod, Massachusetts, to Georgia.

2. S. PYRUM, Dillwyn. Fig. 28.
(*Bulla.*) Desc. Cat., i. 485. 1817.

Pyrula spirata, Lamarck, Anim. s. Vert., vii. 142. 1822.
Fulgur pyruloides, Say. Journ. Philad. Acad., ii. 237.
Busycon plagosum, Conrad, Prov. Acad. Nat. Sciences, Philad. 583. 1862.

Shell with spiral striæ alternately larger; whorls white, transversely lineated with ferruginous lines, interrupted or obsolete on the middle; above flattened, unarmed; spire short, suture profoundly canaliculate.

Southern Coast, Gulf of Mexico.

** *Suture not canaliculate.*

3. S. CARICUS, Gmelin. Figs. 29, 30, 31 (reversed variety).
(*Murex.*) Syst. Nat., 3545. 1788.

Murex aruanus, Linn. (Misnomer), Syst. Nat., edit. xii. 1222. 1766.
Fulgur eliceans, Montfort, Conch. Syst., ii. 503, fig. 1810.
Pyrula carica, Lamarck, An. sans Vert., vii. 138. 1822.
Pyrula candelabrum, Lamarck, ibid.
Pyrula Kieneri, Philippi.
Busyeon spinosum, Conrad, Proc. Acad. Nat. Sci., Philad. 583. 1862.
Busycon gibbosum, Conrad, ibid., 262. 1862.

Shell large, solid, pear-shaped; whorls six, flattened at the summit and the angle raised into a series of compressed tubercles, generally about nine in number, on each volution. Young shell striate within the aperture, striæ becoming obsolete when full grown.

A figure of the egg-case is given for comparison with that of *S. canaliculatus.*

Fig. 31 represents a reversed shell of this species from the southern coast, where it rarely occurs. It is the var. of *perversus* of Kiener, and the *Kieneri* of Philippi, *gibbosus* of Conrad.

Cape Cod, Massachusetts, to Florida.

4. S. PERVERSUS, Linnæus. Fig. 32.

(*Murex.*) Syst. Nat., edit. 12, p. 1222. 1766.

Shell sinistral, pyriform, ventricose, canal elongated; whorls encircled with slightly waved rather distant striæ, angulated round the upper part, coronated round the angle with tubercles indicating slight folds of the surface. Yellowish banded or longitudinally streaked with brown; interior white.

Southern Coast.

Genus FICUS, Klein.

Tent. Method. Ostracol. 1753.

1. S. PAPYRACEUS, Say. Fig. 33.

(*Pyrula.*) Journ. Philad. Acad., ii. 238. 1822.

Shell inflated, thin; spire not elevated; suture slightly impressed, not shouldered, yellowish with small rufous spots; within dull fulvous; whorls with numerous spiral striæ, which are alternately larger, crossed by smaller striæ.

Georgia to West Indies.

Genus FUSUS, Bruguiere.

Encyc. Meth., i. 15. 1789.

Shell fusiform; spire many-whorled; canal straight, long; operculum ovate, curved, nucleus apical.

There are nearly two hundred species; distribution universal.

* *Shell ventricose, rounded.* Nos. 1, 2, 3.
* *Shell angulated by revolving ribs.* Nos. 4, 5.

1. F. ISLANDICUS, Gmelin.   Fig. 34.
   (*Murex.*) Syst. Nat., 3555. 1790.
Murex corneus, Pennant, Brit. Zool., iv. 124, t. 76, f. 99. 1777.
Fusus corneus, Say, Amer. Conch., t. 29.

Shell elongated, symmetrically fusiform; spire regularly attenuated to the apex; volutions eight, slightly convex; body-whorl equally inflated, its surface covered with between forty and fifty small revolving ribs which are conspicuous through the epidermis; these become almost effaced towards the outer lip, when the vertical, sinuous striæ appear in their place. These ribs, or revolving elevated lines, are reduced to fifteen on the next whorl above, diminishing in numbers as they ascend, the intervening furrows becoming more profound, with very faint traces of vertical lines. Aperture oblong-ovate, half the length of the shell; canal short, sinuous and wide. Callus on the columella broad; lip sharp, very minutely impressed by the terminations of the revolving lines. Color— Epidermis horn-colored or soiled brown; surface beneath whitish opalescent; within, pearly white.

Length of shell 2.9 inches.

Animal white, with irregular black spots; foot rounded, rectangular; eyes black.

Massachusetts, northward.

2. F. PYGMÆUS, Gould.   Fig. 35.
   Invert. Mass., 1st edit., 284, f. 199. 1841.

Shell same shape and similarly marked as *F. Islandicus*, but having only six whorls and only one-fourth the size, being about three-quarters of an inch in length. Animal pure white, with large foot, broadly truncate in front.

Maine, northward.

3. F. VENTRICOSUS, Gray.   Fig. 36.
   Zool., Beechy's Voyage, 117.
Fusus Islandicus (Var.), Kiener, Sp. t. 15, f. 2.

Shell subfusiform, ventricose; whorls five, rounded, rapidly attenuating to a blunt apex; body-whorl much inflated, composing the greater part of the shell. Surface covered with a velvety epidermis, under which numerous minute and regular revolving lines, with a few vertical wrinkles, are apparent. Spire short, not exceeding .4 above the body-whorl; lip simple, smooth; columella with a broad callus; canal slightly recurved. Epidermis chestnut-color; beneath, white.

Length nearly 2 inches.

Much more ventricose than *F. Islandicus*, with proportionately larger aperture, and more numerous and smaller revolving lines.

Newfoundland, etc.

4. F. TORNATUS, Gould. Fig. 37.

Am. Journ. Science, xxxviii. 197.

Shell large, coarse, turreted; whorls eight, very convex, rather ventricose, with distant elevated revolving ribs; on the upper whorls, two of these, more prominent than the rest, give them a bicarinated appearance. Suture deep. Incremental striæ distinct, but otherwise the shell has a smooth and worn appearance. Aperture rather less than half the length of the shell, broad-oval, and somewhat dilated; lip sharp, and somewhat angulated by the prominent revolving ribs; in adults the columella margin covered with a callus. Canal short, much recurved. Color faint brownish horn-color; ribs light chestnut-color.

Length 2½, width 1¼ inches.

Newfoundland.

5. F. DECEMCOSTATUS, Say. Fig. 38.

Journ. Acad. Philad., v. 214.

Fusus carinatus, Kiener, Species, t. 19, f. 1.

Shell large, robust, solid, somewhat ventricose, oval; whorls six or seven, obliquely flattened above the shoulder, and with stout, coarse revolving ribs; there are about ten of these ribs on the body-whorl, gradually diminishing below. On the upper whorls, the ribs are reduced to two or three large and coarse ones, which give a turreted appearance to the spire; between these ribs are smaller revolving lines, and the whole surface is coarsely wrinkled by the lines of growth. Aperture ovate; lip festooned by the termination of the revolving ribs; pillar lip arched, and with a broad callus; beak cancellate externally; canal short and curved. Brownish-white or ash-colored; pearly white within, grooves on the lip chestnut-colored.

Length 2 5 inches.

This is the *F. carinatus* of Kiener, but not of Lamarck. It is figured by Reeve (Iconog.) as *Buc. lyratum*, Mart. (*Murex glomus cereus*, Chemn.) from Australia; but Martyn's species is certainly distinct from ours.

Massachusetts, northwards.

Subgenus Trophon, Montfort.

Shell fusiform, varices numerous, lamelliform or laciniated; spire prominent; aperture ovate; canal open, usually turning to the left; columella smooth, arcuated.

These animals inhabit deep water in the Arctic Seas; they are distinguished from the typical Fusidæ by their smaller size and lamellar varices.

1. T. TRUNCATUS, Ström. Fig. 39.

> (*Buccinum.*) Norsk. Vid. Selsk. Skr., iv. 369, t. 16, f. 26.
> *Trophon clathratus*, Forbes & Hanley, Brit. Moll., iii. 436, t. 111, f. 1, 2.
> (not *Murex clathratus*, Linn.)
> *Murex Bamffius* (part.), Donovan, Brit. Shells, v. 169, f. 2.

Shell small, brownish, with six rounded whorls ornamented with numerous sharp lamellar longitudinal ribs; aperture rounded, columella arcuated, canal turned to the left.

Length 13, diam. 6 mill.

> *Maine, northwards ; England.*

2. T. SCALARIFORMIS, Gould. Fig. 40.

> (*Fusus.*) Silliman's Journ. Sci., xxxviii. 197.
> *Murex Bamffius* (part.), Donovan, Brit. Shells, v. t. 169, f. 1.

Shell reddish-brown; whorls seven, convex, with deep sutures; fifteen to twenty blade-like ribs disposed longitudinally on the surface; aperture rounded, canal slightly turned to the left.

Length 43, diam. 20 mill.

> *Massachusetts, northwards.*

Larger, with one more whorl than *T. truncatus*.

3. T. MURICATUS, Montagu. Fig. 41.

> (*Murex.*) Test. Brit. 262, t. 9, f. 2.

Shell small, yellowish or white; whorls seven, very convex; with ten conspicuous ribs crossed by elevated revolving lines, canal long and straight.

Length 17, diam. 7 mill.

> *Massachusetts, Europe.*

Family BUCCINIDÆ.

Shell notched in front; or with the canal abruptly reflected, producing a kind of varix on the front of the shell. Carnivorous.

Synopsis of Genera.

Shell ovate, few-whorled, with a corneous epidermis; whorls ventricose; aperture large, oval, emarginate in front; canal wide, truncated, dorsally more or less tumid; columella smooth, inner lip expanded; outer lip usually thin, smooth internally. Operculum lamellar, nucleus external.

BUCCINUM, Linn.

Shell ovate, small, few-whorled; aperture moderate; columellar lip callous, expanded, forming a tooth-like projection near the anterior canal. Operculum ovate, nucleus apical. NASSA, Lam.

Shell striated, imbricated, or tuberculated; spire short; aperture large, with a slight oblique channel in front; inner lip much worn and flattened. Operculum lamellar, nucleus external. PURPURA, Lam.

Shell very small, limpet like; with a large aperture, channelled in front, and a minute lateral spire. No operculum. PEDICULARIA, Swains.

Shell oblong, small, thick, with a long, narrow aperture; outer lip thickened, especially in the middle, dentated; inner lip crenulated. Operculum very small, lamellar. COLUMBELLA, Lam.

Shell obconic, ventricose, with irregular varices; spire short; aperture long and narrow; outer lip reflected, denticulated; inner lip spread over the body-whorl; canal sharply recurved, producing a varix on the back of the shell. Operculum small, elongated; nucleus in the middle of the straight inner edge. CASSIS, LAM.

Shell oval, ventricose, thin, spirally furrowed, without varices; spire small; aperture very large; outer lip crenated. No operculum.

DOLIUM, Lam.

Shell long turreted, dextral, many-whorled, granulated; aperture subrotund; inner lip broadly reflected; outer lip acute, arcuated and produced anteriorly; aperture anteriorly sinuated; canal short. Operculum corneus, concentric; apex terminal. CERITHIOPSIS, Forbes & Hanley.

Shell long pointed, many-whorled; aperture small; canal short. Operculum pointed, nucleus apical. ACUS, Humphrey.

Shell cylindrical, polished; spire very short, suture channelled; aperture long, narrow, notched in front; columella callous, striated obliquely; body-whorl furrowed near the base. No operculum.

OLIVA, Lam.

Genus BUCCINUM, Linnæus.

Syst. Nat., edit. x. 734. 1758.

Tritonium, O. Fabricius (not Link), Faun. Grœn. 395. 1780.

Shell ovate or oblong, covered by a horny epidermis; spire elevated, apex acute; aperture large, oval, emarginate in front; canal wide, truncated, dorsally more or less tumid; columella smooth; inner lip expanded; outer lip usually thin, smooth internally.

Animal with eyes on slight eminences at the outer bases of the tentacles.

The genus contains few species, of Arctic distribution.

Synopsis of Species.

A. Body-whorl angulated or carinated.   No. 1.
B. Body-whorl not angulated.
  a. Aperture narrow ; a strong, tooth-like plait on the columella.   No. 2.
  b. Aperture broad.
    Shell thick, coarsely striated ; with longitudinal folds.   No. 3.
    Shell thin, finely striated ; with longitudinal folds.   No. 4.
    Shell thin, finely striated ; longitudinal folds obsolete.   No. 5.

1. B. DONOVANI, Gray.   Fig. 43.
    Zool., Beechey's Voy. 128.   1839.
    Stimpson, Review of Northern Buccinums, Canad. Nat.   Oct. 1865.

Buccinum glaciale, Donovan (not Linn.), Brit. Shells, v. t. 154.   1799.
Buccinum tubulosum, Reeve, Icon. f. 105.   1847.

Shell elongated, thick ; spire long and tapering ; whorls 9, convex, with an obtuse carina at the middle of the body-whorl, sometimes obsolete.   This carina commences at the upper angle of the aperture.   Longitudinal folds about thirteen, most distinct near the sutures, and often obsolete on the body-whorl except at the suture.   Primary spiral grooves always double or triple in fresh and good examples.   Primary ridges not much flattened, with fine secondary grooves upon them, but the middle groove is often somewhat deeper than the others.   Aperture about two-fifths as long as the shell, and rounded.   Columellar lip incurved above, and projecting below.   Outer lip somewhat thickened and reflected, patulous, and broadly sinuated above about half-way between the suture and the junction of the carina.   Periostraca very thin and not ciliated.
    Length 68, diam. 53 mill.

Banks of Newfoundland, northwards.

2. B. CILIATUM, O. Fabricius.   Fig. 44.
    (*Tritonium.*)   Fauna, Grœnlandica, 401.   1780.

Buccinum ciliatum (Fab.), Stimpson, Rev. of Buc.   1865.
Buccinum cyaneum, Hancock (not Brug.), Ann. and Mag. Nat. Hist., 1 ser. xviii. 328.   1846.
Buccinum Mölleri, Reeve, Icon. Buc. Errata.   1846.
Tritonium (Bucc.) *tenebrosum,* var. *borealis,* Middendorff (not Hancock), Mal. Ross. 162, t. 3, f. 7, 8.   1849.

Shell rather small and solid, becoming very thick with age, elongated, oval, or sub-elliptical, appressed.   Sutures not impressed ; spire short ; body-whorl elongated, and constituting seven-tenths of the length of the shell.   Whorls not convex, not

carinated, plicated; longitudinal folds thirteen to eighteen in
number, more or less oblique, variable in number and prominence,
but never entirely obsolete at the suture. Primary spiral ridges
narrow and distant, about thirty in number on the lower whorl,
but somewhat variable in strength and distance; sometimes
double or divided in two by a groove. Secondary ridges alternate with the primaries singly or by groups of two, three, or four;
they are distinguished from the primaries by being less prominent,
and occupying the depressions constituting the primary grooves.
In some specimens the primary and secondary ridges and grooves
can scarcely be distinguished from each other. Aperture elliptical,
elongated and narrow, a little more than half the length of the
shell, not patulous, but somewhat canaliculated and projecting
below; outer lip scarcely at all sinuated. Columella with a distinct tooth or projection near its anterior or lower extremity.
This projection corresponds to the second fold of the columella
seen in several species, such as *B. tenue* and *B. undatum*, but it
is more tooth-like than in any other species of the genus, and
constitutes an important and easily recognized specific character.
Periostraca ciliated.

Length 38, diam. 20 mill.

This is not the *ciliatum* of Gould's "Invertebrata of Massachusetts." That species is *B. Humphreysianum*, Bennett.

Nova Scotia, Newfoundland Banks, northwards.

3. B. UNDULATUM, Möller.   Fig. 45.

Kroyer's Tidsskrift, iv. 84.   1842.
Stimpson, Review of Buccinum.   1865.

Buccinum undatum, Gould (not Linn.), Invert. Mass. 305.   1841.
Tritonium undulatum, Mörch, Rink's Greenland. 84.   1857.
Buccinum Labradorense, Reeve, Conch. Icon., f. 5.   1846.

Shell thick, ovate-conic, ventricose, grayish or brownish-white,
encircled by prominent primary raised lines and minute intervening secondary striæ; with twelve or thirteen longitudinal,
obliquely waved, elevated ribs or plaits on the spire and upper
portion of the body-whorl; epidermis velvety, yellowish-brown;
whorls six, regularly convex; aperture oval, about one-half the
length of the shell, golden yellow within; minute striæ extend
some distance within the mouth and produce faint crenulations of
the outer lip; this is somewhat everted and arched so as often to
produce a conspicuous notch at about its posterior third; colu-

mella with a broad callus, somewhat flattened and twisted at its lower portion; not extending so far as the lip on the opposite side of the canal.

Length 75, diam. 47 mill.

New Jersey (rare), northwards.

This is the American prototype of the *B. undatum* of Europe, which it closely resembles. It is distinguished by the following differences: the whorls are more convex, the spire is shorter, and the diameter of shell proportionally greater; the aperture is smaller and more circular; the sinus of the outer lip is broader and shallower, and nearer the middle of the lip; the aperture is (in fresh specimens) golden within, while in *undatum* it is white or chocolate-colored; the columella is shorter; the ciliation of the periostraca is short and sparse, never long and furry as in good specimens of the European species; finally, it never reaches the size or number of whorls of the latter.

4. B. TOTTENI, Stimpson. Fig. 46.
> Review of Northern Buccinums, p. 23. 1865.
> Dawson, Canad. Nat., ii. 415, t. 7, f. 5. 1857.

Buccinum ciliatum (part.), Gould., Invert. Mass. 307. 1841.

Shell of moderate size, white, thin; spire acute; suture impressed; whorls seven, regularly convex, neither carinated nor angulated. Longitudinal folds about twenty-two, very regular, straight, not at all oblique, and about equalling their interspaces in width, becoming obsolete on the body-whorl except occasionally at the suture. Transverse striation sharp and regular, the grooves narrow and deeply cut. The primary ridges are very numerous and crowded, the narrower and less prominent ones usually alternating by threes or fours with the stronger ones. The primary grooves are much narrower than the corresponding ridges. The secondary grooves are few in number, occurring for the most part only on the greater ridges. Aperture rather broad, half as long as the shell; outer lip thin, effuse, projecting below, with its superior sinus very broad and shallow or obsolete; folds of the columella little prominent. Periostraca light-yellowish, short-ciliated with triangular fimbriæ at the intersections of the growth lines with the transverse striæ. Color within the aperture white, or pale-yellowish.

Length 53, diam. 33 mill.

Banks of Newfoundland.

Allied to *B. Humphreysianum*, but differs in its plicated and more convex whorls, deeper transverse sculpture and want of color. It might be taken for a thin and delicate form of *B. undulatum*, but is easily distinguished by the number and straightness of the longitudinal plications of the spire-whorls, the more numerous and sharply-cut transverse ridges, and the wider mouth.

5. B. HUMPHREYSIANUM, Bennett. Fig. 47.

Zool. Journ., London, i. 398, t. 22, upper figures. 1825.
Stimpson, Rev. of Buccinums, Canad. Nat. 1865.
Buccinum ventricosum, Kiener, Iconog. Buc., iv. t. 3, f. 7. 1841.
(not of Lamarck.)
Buccinum ciliatum, Gould, Invert. Mass. 307, f. 209. 1841.
(not of Fabricius.)

Shell rather below the medium size, very thin, translucent, pale-brownish, with fulvous or reddish markings, sometimes obsolete. Spire conic; whorls 7, somewhat flattened above and regularly convex below, so as to be faintly shouldered above the middle. They are neither plicated, carinated, nor angulated, and the surface is much smoother than in most species of the genus. The primary ridges are to be distinguished from the secondaries only at the obsolete angle or shoulder of the whorl, where there are generally two or three small ridges on each side of the more prominent ones and the corresponding sulcus. On the middle and lower part of the body-whorl, where the transverse ridges are for the most part equal in size and strength, and equal to the intervening grooves, the latter are crossed by well-marked though microscopic lines of growth. The aperture is almost one-half the length of the shell and about three-fifths as broad as long. The outer lip is very little thickened and scarcely at all projecting below, and it has no sinus at the middle. Periostraca ciliated.

Length 37, diam. 20 mill.

Gulf of St. Lawrence, northwards; Northern Europe.

It may be recognized by its thin structure, superior flattened whorls, and the total absence of plications.

Genus NASSA, Martini.

Verzeich. c. Ausserl. Samml. 1773.

Shell ovate, ventricose; body-whorl variously sculptured; aperture ovate, with a short, reflected, truncated anterior canal; inner lip smooth, often widely spread over with enamel, with a posterior callosity or blunt dentiform plait; outer lip dentated, internally crenulated.   Operculum ovate, the margin serrated or entire. Eyes on the middle of the tentacles.   Foot large, expanded, frequently bifurcate at its posterior extremity.

The animal is exceedingly active in its movements, feeding on bivalves, which it pierces with its proboscis, extracting the contents through a small, round aperture.   There are over two hundred species, of world-wide distribution; ranging from low water to fifty fathoms.

1. N. OBSOLETA, Say.   Figs. 48, 49, and 50.
     Journ. Philad. Acad. Nat. Sciences, ii. 232.   1822.

Ilyanassa obsoleta, Stimpson, Am. Journ. Conch., i. 61.   1865.
Buccinum Nov-Eboracensis, Wood, Index Test, Suppl., t. 4, f. 26.
Buccinum oliviforme, Kiener, Iconog., t. 25, f. 99.

Shell ovate, solid; apex eroded; spire short; whorls six, flattened, convex, reticulated, the ridges flattened; aperture rather less than half the total length of the shell, outer lip simple, striated within.   Dark reddish-brown, purplish within, with frequently a white revolving band.   Operculum obovate, broadest below; nucleus a little within the margin at the outer side near the base; margin entire, not serrated.

Animal variously mottled with slate-color; foot as long as the shell, its anterior angles prolonged and turned backwards, and without caudal bifurcation ; head not extending beyond the shell; eyes black, on the exterior side of the tentacula, and above the base ; above the eyes the tentacula are suddenly diminished and bristle-shaped ; proboscis cylindrical, half the length of the shell, channelled beneath.

Ova-capsules (fig. 50) rounded, erect, slightly compressed, with the anterior and upper surface covered with facettes formed by reticulating ridges or crests, the angles of which are spinous.

This is the type of Stimpson's genus *Ilyanasa*, differing from the typical *Nassæ* in the form of the operculum and the want of caudal bifurcation.

Littoral, living on mud-flats in bays and harbors; very abundant.
                                   Entire Coast from Maine to Florida.
2. N. TRIVITTATA, Say.   Fig. 51.
     Journ. Philad. Acad. Nat. Sc., ii. 231.   1822.

Shell robust, ovate-conic; spire elevated, acute, longer than the

3

body-whorl. Whorls six or seven, flattened; surface granulated by prominent vertical lines and about ten revolving impressed lines. Suture impressed, with a prominent shoulder on the whorl near it. Aperture oval; lip sharp, sculptured with revolving striæ; columella, with a slight fold, white; with, unfrequently, three brownish bands. Operculum subtriangular, dentate around the margin.

Length 13 to 19 mill., diam. 6 to 8 mill.

The animal is whitish with light purple dots, and its foot is bifurcate behind. Very common from low-water mark to fifteen fathoms.

New England to Georgia.

3. N. VIBEX, Say.   Figs. 52, 53.

Journ. Philad. Acad. Nat. Sc., ii. 231.   1822.

Nassa fretensis, Perkins, Bost. Proc., 117.   1869.

Shell solid, ovate, short; whorls six; body-whorl with from ten to twelve vertical, undulating and prominent costæ, which are continued to the apex; and about the same number of revolving lines, which are most prominent on the costæ; suture moderate. Aperture oval; lip thickened without and within, with two to four prominent teeth internally; pillar lip arched, with a broad flat callus, which forms a process directed upwards towards the suture on the upper portion of the body-whorl, and is slightly granulated at the base. Spire short, rapidly attenuated to an acute apex; canal very short. Color ashy-white to pale reddish-brown, with darker colored revolving bands.

Length 13, diam. 8 mill.

The animal has a large foot, auriculate in front, and narrowed behind, about one-half longer than the shell. The broad head is maculated with dark gray, and the upper part of the body with snow-white, and a broad longitudinal median band of the same color; beneath, whitish.

A rather larger variety, with less acute spire, more numerous and narrower ribs, more prominent revolving lines, thinner and smaller callus and darker color, has been recently described as *N. fretensis.* I give a copy of the figure (fig. 53). From low water to fifteen fathoms.

Cape Cod, Massachusetts, to West Indies.

4. N. ACUTA, Say. Fig. 54.
>Journ. Philad. Acad., ii. 234. 1822.

Conic-acute, cancellate, so as to appear granulate; granules prominent, somewhat transverse, inequidistant. Spiral grooves six in number; spire longer than the body-whorl, slender, acute. Beak distinguished by a depression from the body-whorl, and slightly reflected; lip thickened, with elevated lines on the fauces, not attaining the margin. Color whitish.

Length 13 mill.

Southern Coast.

5. N. UNICINCTA, Say. Fig. 55.
>Journ. Philad. Acad., v. 211. 1826.

Shell sub-ovate, conical. Whorls eight, with ten to twelve revolving lines and transverse undulations; apex acute; lip with ten revolving striæ within; pillar lip concave in the middle; two obsolete striæ, and a deeper one at the base. Color yellowish-white or ash-gray; body-whorl with a brown band.

Length 23 mill.

South Carolina.

6. N. CONSENSA, Ravenel.
>Proc. Acad. Nat. Sciences, Philad. 43. 1861.

Shell ovate-conical, ribbed and crossed by numerous revolving striæ; whorls seven, and the apex; whorls rounded, with eleven strong ribs; suture deep, scalloped by the ribs; revolving striæ crossing the ribs, as well as the interstitial spaces.

Aperture nearly oval, outer lip much thickened, denticulate within, the largest tooth being in the form of a ridge next the canal; pillar much hollowed, with slight callus above, much thickened to form the canal, which is short, oblique, and turned backwards; lower portion of the pillar white, covered with crowded, inconspicuous, revolving striæ, with a deep groove at the edge of the canal.

Color of the shell generally yellowish-brown, with a narrow deep-brown band immediately next the white projection at the canal; next to this, on the body-whorl, is a much wider band of lighter brown, which revolves at the suture to the apex of the shell; all other portions of the surface are marked by delicate lines, more or less grouped, of yellowish-brown. On the thickened portion of the outer lip these lines are here and there more deeply colored in spots.

It is a very pretty shell ; a single specimen was found in a fish off Charleston bar ; fourteen fathoms. It resembles *N. incrassata* of England, and we have seen it in collections labelled " *N. ambigua*, Mor., West Indies."

This species I have not been able to identify. It has never been figured, and, as only a single specimen was obtained, it may be considered a doubtful species.

Genus PURPURA, Bruguiere.
Encyc. Method., i. 15, 241. 1789.

Shell oblong-oval, last whorl large ; spire short ; aperture ovate, large, with an oblique channel or groove at the forepart ; columella flattened ; outer lip simple.

There are about one hundred and fifty species ; inhabiting the seas of temperate and warm climates.

1. P. LAPILLUS, Linnæus. Figs. 56, 57.
(*Buccinum.*) Syst. Nat., edit. xii. 1202.

Shell ovate, thick and solid ; spire short and very acute, suture impressed ; whorls five, with deep revolving furrows and intervening ribs, and transverse raised scale-like wrinkles. Aperture ovate ; lip arched and subacute, with obscure revolving ridges within the margin. Pillar lip produced, concave externally at the base ; canal short. Operculum horny, oval. Color from white, through various shades of yellow, to dark red, sometimes with a white band.

Length 31, diam. 17 mill.

Inhabits rocky shores. It is a very variable species in coloration and form, and is equally common on both sides of the Atlantic. ·
Entire Coast, Greenland to Florida. (Europe.)

2. P. FLORIDANA, Conrad. Fig. 58.
Journ. Philad. Acad. Nat. Sci., vii. t. 20 f. 21. 1837.

Shell oblong-ovate, conspicuously attenuated at both ends, spire sharp ; whorls depressed and finely noduled around the upper part, transversely very closely ridged and grooved throughout, interior of the aperture deeply grooved. Bluish-gray, indistinctly and irregularly encircled with narrow yellow zones, and rather indistinctly painted with blackish longitudinal waves, columella and interior of the aperture orange-yellow, frequently banded within.
North Carolina to Florida.

Genus **PEDICULARIA**, Swainson.
Man. Malacol. 245, 357. 1840.

Shell very small, limpet-like; with a large aperture, channelled in front, and a minute lateral spire. The following species is the second in the genus, the typical species being a small shell parasitic on corals in the Mediterranean; in the latter there is no operculum. The genus is closely allied to *Purpura*.

1. P. DECUSSATA, Gould.
Proc. Bost. Soc. Nat. Hist., v. 126. 1855.

Shell solid, variable, generally oval, decussated by radiating and revolving striæ; apex lateral, obtuse; aperture crescent form; lip thickened, considerably expanded; columella straight, acute, sulcate.

Length 4, diam. 3 mill.

Dredged at a depth of 400 fathoms.

Coast of Georgia.

Genus **COLUMBELLA**, Lamarck.
Anim. sans Vert. vii. 292. 1822.

Shell ovate-oblong, triangular or fusiform; spire acute at the apex; aperture long, narrow, contracted in the middle; inner lip curved, crenulated or denticulated; outer lip dentate, gibbous, thickened in the middle.

There are nearly two hundred species of this genus, inhabiting temperate and warm seas. They crawl on the surface of sand-flats in shallow water, or live on stony beaches, where they congregate about and under stones.

1. C. AVARA, Say. Fig. 59.
Journ. Philad. Acad., ii. 230. 1822.

Shell thick, small, elongate-ovate; spire elevated and acute. Whorls six or seven, very slightly convex, almost flat; suture distinct. Surface with spiral impressed lines, and vertical obtuse ribs or folds; these latter, consisting of about twelve or fourteen in number on the body-whorl, do not descend, beyond the middle of that whorl, leaving only revolving lines beneath. Columella with a plate of enamel, which is toothed within and truncated beneath the margin; lip toothed within. Color whitish, reticulated or spotted with rufous; often of a yellowish hue.

Length 12, diam. 4 mill.

Cape Cod, Mass., to Georgia.

2. C. ROSACEA, Gould.   Fig. 60.
    Am. Journ. Sci., xxxviii. 197.   1840.

C. Holböllii, Beck. Moll.   1842.

Shell small, elongated, closely covered by very minute revolving lines; spire acute, about as long as the body-whorl, the suture distinct, but faintly impressed; aperture narrow; outer lip sharp, smooth within; columella arcuated, smooth, a little flattened. White tinged with rose-color, whitish within the aperture.

Length 7.5, diam. 3.75 mill.
            Massachusetts northward to Newfoundland.   (*Europe.*)

· 3. C. LUNATA, Say.   Fig. 61.
    Journ. Philad. Acad., v. 213.   1826.

C. Gouldiana, Agassiz, Stimpson, Shells of N. England, 48.   1851.

Shell very small, conic-oval. Whorls six, nearly smooth, slightly convex: a single revolving line below the suture, and a few around the base; suture not deeply impressed. Aperture narrow, slightly angulated above, and with a short channel beneath. Columella with a callus; lip simple, dentate on its inner margin; those above most prominent. Color reddish-brown or yellowish, with one or more series of sublunate white spots on the body-whorl; occasionally uniform reddish-brown, or with longitudinal dark lines.

Length 5, diam. 2.5 mill.

Animal pale whitish; the foot linear, nearly as long as the shell, acute behind, truncate before; proboscis more than half the length of the shell, obtuse at tip with a brown annulation and another at the base; tentacula short, cylindrical, annulate with blackish on the middle; eyes black, at the base of the tentacula.

                    Cape Cod, Mass., to Georgia.

4. C. WHEATLEYI, DeKay.   Fig. 62.
    (*Buccinum.*)   Moll. N. York, 132, t. 7, f. 162.   1843.

Shell minute, small, ovate-cylindrical. Whorls six, nearly flat, or at most very slightly convex, with a small and distinct suture; surface smooth, with no revolving lines. Aperture narrow, sub-linear with a small notch above and a short canal beneath. Body-whorl, on its lower portion, near the canal, has from eight to ten minute impressed revolving striæ, becoming more distant above. Lip simple, thin, with a ridge of minute teeth within its inner edge, which are entirely wanting in the young. Callus on the columella elevated, not much reflected.

                                                    /

Color light horn, with numerous undulated vertical reddish dilated lines.

Length 0.23 ; of aperture, 0.1.

New York Harbor.

The above is a copy of DeKay's description. The species has not occurred to any other collector, and it is therefore impossible to decide upon its distinctness. DeKay himself, in describing *C. lunata*, suggests that this species is a variety of it, and the same opinion has been advanced by several marine conchologists.

5. C. DISSIMILIS, Stimpson. Fig. 63.

> Proc. Bost. Soc. Nat. Hist., iv. 114. 1851.

Buccinum zonale, Linsley, Am. Journ. Science, O. S. xlviii. 285.

Shell small, ovate-conical, solid, longitudinally sub-striate, fuscous, often with three white zones; whorls five, flattened; aperture small, sub-equalling half the spire. Animal white.

Length 5, diam. 2.5 mill.

New England.

6. C. SIMILIS, Ravenel. Fig. 64.

> Proc. Acad. Nat. Sciences, Philad., 41. 1861.

This has generally been considered the young or immature shell of *C. avara*. The latter is a larger shell, and has fewer and much larger ribs on the upper portion of the body-whorl. It has about eleven ribs; whereas this shell has often as many as twenty ribs, but the number varies; the ribs are also smaller and more regular generally, occupy more of the whorl, and generally are continued on the spire to its apex. White mottled with rufous.

The principal distinction between this species and *C. avara* consists in its narrower form and generally more numerous ribs. Although frequently confounded with the latter—with which it inhabits, it appears to be quite distinct.

Massachusetts to Georgia.

7. C. TRANSLIRATA, Ravenel.

> Proc. Philad. Acad. Nat. Sci., 42. 1861.

Shell elevated conic, apex sharp; whorls nine, nearly flat, rather closely ribbed, ribs and interspaces about equal, with five equidistant revolving striæ, from the anterior canal to the apex; body-whorl angulated in the middle, and above this angle the ribs are about half as many as on the whorl immediately preceding it and nodulous at the suture; the nodules being white. Below the angle the ribs are again more numerous but not so prominent, with numerous revolving striæ. Color varying from light straw-yellow to dark-brown, the ends of the ribs at the suture and on the angle of the body-whorl, white.

Length one inch.

Beaufort, N. C., Charleston, S. C.

I am not acquainted with this species.

8. C. IONTHA, Ravenel.

Proc. Philad. Acad. Nat. Sci., 42: 1861.

Shell fusiform, strong, small, with nine flat, ribbed whorls, white with brown blotches and lines. Suture deep and distinct, both the upper and lower edges of the whorls being chamfered; the ribs on the body-whorl near the aperture less distinct than on other parts of the shell; anterior portion of this whorl with numerous revolving striæ; these impressed striæ give place to colored lines as they ascend, and these are continued more or less distinct to the apex, being visible only as they cross the ribs and not in the intermediate spaces except here and there, where, being more deeply colored and descending between the ribs, they produce the blotches which mark the shell. Aperture small, rather wide in proportion, pillar lip much hollowed above, suddenly becoming straight to form the canal; outer lip considerably enlarged, denticulated sparsely within; length a little over one-fourth of an inch.

A pretty little shell, allied to the group which embraces C. pulchella, Sowb. and C. jaspidea, Sowb. of West Indies; this species being more elongated.

A single specimen obtained from the stomach of a black fish.

Charleston, S. C.

9. C. NIVEA, Ravenel.

Proc. Philad. Acad. Nat. Sci., 43. 1861.

Shell small, delicate, elongated-conic, white, immaculate, smooth, polished, prettily striated on the outer part of the canal, body-whorl longer than the spire, suture distinct, with a white revolving line a little below it on the whorls; pillar covered with callus, much hollowed, suddenly becoming straight to form the canal; callus ending in a distinct edge; outer lip a little thickened, sparsely denticulated within, the posterior tooth being decidedly the most prominent.

Allied to *rosacea*, Gould, and *lunata*, Say. A single specimen taken from the stomach of a fish.

Charleston, S. C.

This, as well as *C. translirata* and *C. iontha*, is a doubtful species.

10. C. SPIZANTHA, Ravenel.

Proc. Elliott Soc. Nat. Hist., 281. 1858.

Shell small, ovate-conic; smooth, except at the base, where there are a few revolving lines; whorls seven in mature specimens; nearly flat, with the suture distinct; color brown, with a series of irregular triangular spots, of a dull yellow. There is considerable variation in the coloring, sometimes the general color is of the dull yellow, with brown, waving lines, marking off the whorls with the irregular spots. Aperture oval, about one-third the length of the shell, with a slight recess at the posterior angle, and a short canal in front; brown, with a few teeth within the outer lip and a smooth slight callus on the pillar. Length about one-sixth of an inch.

Animal white; proboscis half the length of the shell, foot a little longer than the shell, narrow, wider in front; posterior end quite narrow, but not pointed, head projecting from the foot, with tentacles one-third the length of the shell, very delicate almost hair-like, with small black eyes at the base. Animal active, keeping the proboscis in constant motion, while the tentacles are little used.

Wando River, S. Carolina.

This species is unknown to me.

11. C. MERCATORIA, Linnæus. Fig. 65, 66.

Syst. Nat. Edit. xii. 1190. 1767.

Shell obovate, thick, spirally grooved throughout, body-whorl swollen, obtusely shouldered, contracted below; whorls six; spire short, conic, apex acute; aperture long and narrow, sinuous, the outer lip thickened, very callously denticulated in the middle. Shell variously colored with longitudinal blotches of brown on a white ground, sometimes with white bands bearing black articulations.

This well-known species is chiefly characterized by its somewhat tumid growth and grooved sculpture, and by its painting, which is generally sparingly articulated.

North Carolina to West Indies.

12. C. ORNATA, Ravenel.   Fig. 66, a.
  Proc. Elliott Soc. Nat. Hist., i. 281.   1858.

Shell small, dirty white, ovate conic; whorls seven, nearly flat, with longitudinal ribs extending almost to the apex; revolving lines interrupted at the ribs, except near the base, where the ribs become obsolete, and the revolving lines are uninterrupted and more decided than elsewhere; suture distinct, with the revolving line next below it more deeply impressed than the others; aperture nearly half the length of the shell, narrow, with a rather deep sinus at its posterior angle, ending in a short canal in front; outer lip thickened and smooth on the outside, being free from the ribs and lines of the whorls, within strongly toothed; pillar covered with smooth callus, the outer edge of which is elevated and sharp.

Length 5 mill.

South Carolina.

I have not seen this species.

Genus CASSIS, Lamarck.
Ann. du Mus., ii. 168.   1801.

Shell obtriangular or obovate; spire short; last whorl large, with irregular varices: aperture linear, long, with a short, sharply recurved, sinistral canal in front; inner lip forming a large, transversely wrinkled plate spread over the body-whorl; outer lip thickened, reflected, plicate or toothed.

The Cassides are active and voracious, living in sandy localities where bivalves abound, and upon which they prey. There are about forty species, principally tropical.

1. C. CAMEO, Stimpson.   Fig. 67.
  Am. Journ. Science, n. s., xxix. 443.   1860.
Cassis Madagascariensis (part.) Lam. Anim. sans Vert., vii. 219.   1822.

Shell ovately triangular, ventricose, irregularly coarsely striate by elevated growth lines, more prominent on the spire, encircled by rounded riblike elevations, of which, on the body-whorl, three are elevated occasionally into nodules, the most prominent being the upper one, which forms the superior angle of the body-whorl; the spire is very short and sutures not deep; between the revolving ribs are grooves of equal width. Columella greatly expanded, conspicuously ribbed; outer lip expanded and thickened with lamelliform teeth within. Flesh colored, the lips and callus of deeper shade, teeth and wrinkles of mouth white, interstices and interior deep purple-brown.

Length 6, diam. 4 inches.

North Carolina to West Indies.

2. C. GRANULOSA, Bruguire. Fig. 68.
Encyc. Meth., No. 5. 1792.

Shell somewhat globosely ovate, rather thin, extremity ventricose, spire sharp; whorls rounded, inflated, smooth, transversely obsoletely grooved, reticulately striated towards the apex; columella smoothly plicated and conspicuously granose; lip reflected, thickened and dentate within. Bluish-white, encircled by five bands of somewhat square and irregularly interrupted reddish-brown spots.

Length 4, diam. 2¾ inches.

North Carolina to West Indies.

Genus DOLIUM, Lamarck.
Philos. Zool. 1809.

Shell thin, roundly oval, ventricose, inflated; spire small; whorls transversely furrowed; aperture very large, with a short, posterior reflected canal; inner lip thin, widely expanded; outer lip fimbriated or crenated. No operculum.

About twenty species; distribution tropical.

1. D. GALEA, Linnæus. Fig. 69.
(*Buccinum.*) Syst. Nat., edit. xii. 1197. 1767.

Shell very large, extremely ventricose, umbilicated; spire short, a little sunk in; sutures deeply impressed; whorls seven in number, swollen around the upper part, with revolving ribs; ribs convex, close-set, upper ones with most frequently an intervening ridge; columella somewhat twisted; whitish or pale fulvous, apex blackish, columellar lip white, lip stained with brown.

Length 8, diam. 6 inches.

North Carolina to W. Indies ; Mediterranean.

2. D. PERDIX, Linnæus. Fig. 70.
(*Buccinum.*) Syst. Nat., edit. xii. 1197. 1767.

Dolium plumatum, Green, Trans. Albany Inst., i. 131. 1830.

Shell ovately oblong, thin, inflated, obliquely effused towards the base, umbilicated; spire somewhat exserted; whorls six in number with revolving ribs and grooves; ribs about twenty in number, contiguous, flatly convex, scarcely raised; columella arched, lip simple; fulvous brown or light-reddish bay, marked, more or less plentifully, with white lunate spots, interior of the aperture bluish-white or brownish.

Length $6\frac{1}{2}$, diam. $4\frac{1}{2}$ inches.

Southern Coast, W. Indies.

Genus CERITHIOPSIS, Forbes and Hanley.
Brit. Moll., iii. 364. 1853.

Shell turreted, many-whorled, dextral, granulated; aperture subrotund; inner lips broadly reflected; outer lip acute, arcuated and produced anteriorly; aperture anteriorly sinuated; canal short. Operculum corneous, concentric, nucleus terminal.

1. C. EMERSONII, Adams. Fig. 71.

(*Cerithium.*) Bost. Journ. Nat. Hist., ii. 284, t. 4, f. 10.

Shell long, conical; whorls seventeen, flat, each with three rows of granules; suture very deeply impressed; aperture small, subquadrate, about one-sixth the length of the shell; columella spirally twisted; canal less than half the length of the aperture. Color dark reddish-brown.

Length 13, diam. 3 mill.

Animal with a long small foot, truncate in front, and notched behind, pale with white flecks; head broadly rounded, dark flesh-color in front.

New England.

2. C. TEREBRALIS, Adams. Fig. 72.

Bost. Journ. Nat. Hist., iii. 320, t. 3, f. 7. 1841.

Elongated; whorls ten to twelve, flattened, with three or four elevated revolving ridges on each, with numerous fine longitudinal lines between the ridges. Base abrupt; aperture oval, about one-eighth of the total length of the shell. Color reddish-brown.

Length 12, diam. 3 mill.

Animal whitish, with flakes of opaque-white, tentacles clouded

Massachusetts. (*Eur.*)

According to Gwyn Jeffreys this is a synonym of *trilineata*, Phil. (1836).

Genus ACUS, Humphrey.
Mus. Calonn. 1797.

Terebra (part.). Lamarck, Prodr. p. 71. 1799.

Shell subulate, solid, porcellanous; whorls numerous, simple; aperture small, emarginate anteriorly, not produced into a canal;

columella simple, incurved, not tortuous; outer lip simple, acute, without a sinus at the forepart. Operculum annular, nucleus apical. Tentacles very short, with eyes at their tips. Mantle inclosed, with an elongated siphon. Foot small.

1. A. DISLOCATUS, Say. Fig. 73.
    Cerithium. Journ. Philad. Acad., ii. 235. 1822.
Terebro Petitii. Kiener, Spec. Gen. 37, t. 13, f. 32.

Shell small, polished, attenuated; whorls with numerous, minute, impressed revolving lines, and fifteen to eighteen transverse ribs to each whorl, which are dislocated or interrupted near the summit of each whorl by a revolving groove as deep as the suture; color chocolate-brown with a pale revolving band, ribs white.

Length 44, diam. 8 mill.

Maryland, southward.

2. A. CONCAVUS, Say.
    (*Turritella.*) Journ. Philad. Acad., v. 207. 1826.

Shell subulate, white; volutions more than ten, concave in the middle, and sculptured with from two to four obsolete, impressed revolving lines, and with an apical and basal band of about fifteen longitudinal undulations on each volution; the basal band passes round the middle of the body-whorl; suture very slightly impressed, interrupting the continuity of the undulations in the adjacent bands; canal rather prominent.

Length 13, diam. maj. 3 mill.

South Carolina.

Genus **OLIVA**, Bruguière.
Encyc. Meth., i. 15. 1789.

Shell cylindrical, polished; spire very short, suture channelled; aperture long, narrow, notched in front; columella callous, striated obliquely; body-whorl furrowed near the base. No operculum in the typical species.

Animal with a very large foot, in which the shell is half immersed; mantle-lobes large, meeting over the back of the shell, and giving off filaments which lie in the suture and furrow. The eyes are placed near the tips of the tentacles.

About one hundred and twenty species; subtropical.

1. O. LITERATA, Lamarck.　Fig. 74.

　　Ann. du Mus., xvi. 315.　1811.

Say, Am. Couch., i. pl. 3.　1830.

Shell with a pale yellowish-white ground color, thickly covered with cinereous-rufous angulated lines, leaving distinct triangles of the ground color; on each side of the middle is a broad band, occasioned by the angulated lines being there of a deeper or chestnut-brown color; the angulated lines at the upper edge of the volutions are fasciculated and of the same color as the bands.

　　Length 50, diam. 17 mill.

Southern Coast.

A rather large shell, with longer spire, of lighter texture, and more cylindrical than *O. reticularis* of the West Indies, which it much resembles.

2. O. MUTICA, Say.　Figs. 75, 76.

　　Journ. Philad. Acad., ii. 228.　1822.

Shell suboval, white, or yellowish-white; body-whorl with about three revolving maculated bands of pale rufous, of which the superior one is continued upon the spire, the intermediate one is dilated so as to be sometimes confluent with the inferior one, which is the narrowest; spire short; suture very narrow; columella destitute of striæ.

　　Length 10 to 15, diam. 5 to 7 mill.

North Carolina, southwards.

A very common species, generally of the smallest dimensions stated.　In coloring it is infinitely variable, the ground color running through all shades from white to dark chocolate, with or without bands.

Family CONIDÆ.

Shell inversely conical or fusiform; aperture long and narrow, outer lip notched at or near the suture; columella without plaits. Operculum minute, lamellar.

Synopsis of Genera.

Shell turreted, fusiform; spire elevated; aperture oval; *canal long and straight;* columella smooth; outer lip notched anteriorly and with a deep slit near the suture.　Operculum ovate, nucleus apical.

PLEUROTOMA, Lamarck.

Shell turreted; spire raised; aperture oval; *canal short,* recurved; inner lip thickened; outer lip inflexed, with a deep posterior sinus, and a small sinus at the forepart.　DRILLIA, Gray.

Shell ovate, fusiform, small; surface dull, smooth or longitudinally ribbed; spire elevated, shorter than the body-whorl; columella flattened; canal short; outer lip with a small sinus at its junction with the body-whorl. (This genus may be known by its flattened columella.) Operculum as in *Pleurotoma* (Boreal).   BELA, Leach.

Shell turreted, apical whorls nearly smooth, the others ribbed. Posterior sinus of lip very large. No operculum. Animal without eyes.
                                                   PLEUROTOMELLA, Verrill.

Shell solid, subfusiform, smooth or longitudinally ribbed; aperture linear, *with scarcely any canal in front;* columella smooth, simple; outer lip acute, with a slight sinus posteriorly, near the suture. No operculum. (Tropical.)   MANGELIA, Leach.

Genus PLEUROTOMA, Lamarck.

Prodromus. 1799.

1. P. BICARINATA, Couthouy.   Fig. 77.

Bost. Journ. Nat. Hist., ii. 104, t. 1, f. 11. 1838.

Shell minute, tapering at both extremities, turreted; whorls six, convex, with numerous revolving ribs and smaller ones intervening; about the middle a deep groove, with a prominent revolving rib on each side; sutures clearly defined. Aperture narrow, elliptical, ending in a short canal slightly inclining to the left; lip thin, toothed by the revolving ribs, with a slight notch above; pillar-lip arched at its upper third. Whitish, slate-color or dusky-brown.

Length 7.5, diam. 3.75 mill.

Massachusetts (*rare*), *Eur.*

2. P. BRUNNEA, Perkins.   Fig. 78.

Bost. Proc. 121. 1869.

P. plicata, Adams (not Lamarck), Bost. Journ. Nat. Hist., iii. 318, t. 3, f. 6. 1840.

P. declivis, Lovén, 1846.   (Jeffreys.)

Shell minute, thick, fusiform; whorls six, convex; body-whorl with about twelve prominent oblique folds, crossed by ten or more revolving threads, rendering the folds somewhat nodulous; suture deeply impressed; whorls above with folds and revolving lines; spire pointed, somewhat turreted. Aperture narrow, less than half the length of the shell; lip arched, thickened by one of the folds; notch above deep, distinct and smooth. Canal short. Epidermis ashen-brown; beneath this, white; lip brownish within.

Length 6, diam. 3 mill.

Massachusetts, Connecticut, New York (*rare*).

Prof. Adams' name being preoccupied by Lamarck, Mr. Perkins has changed it to *brunnea*, as above.

3. P. CERINA, Kurtz and Stimpson.   Fig. 79.
> Proc. Bost. Soc. Nat. Hist., iv. 115. 1851.

Shell fusiformly turreted, waxy or cinereous, with about ten longitudinal elevated plicæ, and numerous transverse striæ; whorls seven, flattened; aperture oblong, about half equal to the spire; lip simple.

Length 7.5, diam. 2.25 mill.

> *Massachusetts ; Connecticut ; North and South Carolina.*

Genus DRILLIA, Gray.
Jardino's Ann. Nat. Hist., i. 28.   1838.

1. D. ELOZANTHA, Ravenel.
> Proc. Philad. Acad., 43. 1861.

Shell robust, conic cylindrical, with ten whorls, which are bicarinate by being deeply grooved immediately above the suture and again in the upper half of the whorl. The ridge left between these grooves is ornamented by ten strong, yellow, smooth, shining tubercles; the upper edge of the whorl is again bevelled, forming the second carina, which is not at all nodulous. Below the nodulous carina on the body-whorl there are obsolete ribs, crossed by four nodulous ridges, and below these there are eight others, some of which are obscurely nodulous; on all parts of the shell not occupied by the tubercles or carina there are numerous fine equidistant striæ, requiring the glass to bring them to view. Aperture small, outer line made oblique by the deep sinus of the thick outer lip just below the suture; pillar nearly straight, with a strong callus. Color deep brown, with a lighter colored band near the extremity of the canal; beyond that, to the extremity is again almost black. Allied to *P. ornata*, Orb.

Three specimens obtained from stomachs of fish. (Ravenel.)

> *Charleston, S. C.*

Genus BELA, Leach.
Gray, Ann. xx.   1847.

1. B. TURRICULA, Montagu.   Fig. 80.
> (*Murex*) Test. Brit. 262, t. 9, f. 1.   1803.

B. nobilis et *scalaris*, Möller.

B. Americana, Packard.

B. angulata, Reeve.

B. exarata, Müller (Mörch.)

Shell turriculated, thin, with seven or eight shouldered whorls, apex acute; surface with twelve to fourteen oblique, compressed ribs, becoming obsolete below the middle of the body-whorl, and numerous, elevated, spiral lines; outer lip sharp, canal short, open, columella smooth, arcuate. White or yellowish.

Length 16, diam. 6 mill.

Massachusetts and northwards ; England.

Differs from the next species by being broader, more delicate, and by its angulately shouldered whorls.

2. B. HARPULARIA, Couthouy. Fig. 81.

(*Fusus.*) Bost. Journ. Nat. Hist., ii. 106, t. 1, f. 10. 1838.

Shell ovate-oblong; whorls six to eight, round-shouldered, with about eighteen oblique rounded ribs, crossed by fine revolving lines, ribs becoming obsolete on lower part of body-whorl. Brownish flesh color. Aperture narrow, outer lip sharp, inner lip white, smooth, moderately arched, canal very short.

Length 12, diam. 6 mill.

Massachusetts, northward. (Eur.)

It is constant in its color, which differs from both the preceding and following species.

3. B. PLEUROTOMARIA, Couthouy. Fig. 82.

(*Fusus.*) Bost. Journ. Nat. Hist., ii. 107, t. 1, f. 9. 1838.
Fusus rufus, Gould (non auct.), Invert. Mass. 1st edit. 290. 1841.
B. pyramidalis, Ström.
B. Vahlii, Beck.

Shell elongated-fusiform, apex acute; whorls eight, slightly convex, with eighteen to twenty undulating longitudinal ribs and equal interstices covered by very fine spiral striæ; body-whorl more than half the length of the shell, slightly shouldered, the ribs obsolete, but spiral striæ more conspicuous on the lower portion; aperture small and narrow, canal very short; reddish-fawn colored.

Length 12, diam. 3.25 mill.

Cape Cod, Mass., northward.

4. B. CANCELLATA, Mighels and Adams. Fig. 83.

(*Fusus.*) Bost. Journ. Nat. Hist., iv. 52, t. 4, f. 18. 1842.

Shell turreted, slender; with seven convex whorls, decussated by twenty longitudinal ribs crossed by numerous raised revolving lines, giving the surface a cancellated appearance; spire acumin-

4

ate, apex acute; aperture small, narrow, the outer lip crenated by the revolving striæ; white, or tinged with purple.

Length 9, diam. 4 mill.

Casco Bay, Maine (rare), northwards.

5. B. TREVELYANA, Turton. Fig. 84.

(*Pleurotoma.*) Mag. Nat. Hist., vii. 351. 1834.

Pleurotoma decussata, Couthouy. Bost. Journ. Nat. Hist., ii. 183, t. 4, f. 8. 1839.

Shell ovate-fusiform; whorls five or six, convex, with twenty-five or more minute longitudinal ribs or folds, and numerous fine revolving striæ, suture well marked; body-whorl two-thirds the length of the shell, the ribs becoming obsolete on its lower part; aperture narrow, oval, one-half the length of the shell; white or flesh color.

Length 10, diam. 5 mill.

Massachusetts, northward.

6. B. VIOLACEA, Mighels and Adams. Fig. 85.

(*Pleurotoma.*) Bost. Journ. Nat. Hist., iv. 51, t. 4, f. 21. 1842.

Defrancia Beckii. Möll. 1842. (Jeffreys.)

Shell ovate-fusiform; whorls six, convex, longitudinally slightly plicated, spirally finely striate; the body-whorl slightly shouldered and folds becoming obsolete at its middle, spire-whorls with a median revolving carina; aperture small and narrow, canal very short and wide. Dark purple, under a pale-brown epidermis.

Length 7.5, diam. 3 mill.

Massachusetts Bay, northward (Eur.).

7. B. UNDATELLA, Gould.

Proc. Bost. Soc. Nat. Hist., viii. 281. 1862.

Shell ovately rhomboidal, transparent, white, shining, slightly spirally striate; apex mamillate; whorls four tabulate (angulately rounded) with a subsutural impressed line and about eight opaque longitudinal undulations; aperture about half the length of the shell.

Length 3, diam. 1.5 mill.

The transparency and mamillated tip of the minute shell renders it plausible that it may be the young of some larger species. Dredged in 400 fathoms.

Coast of Georgia.

Genus **PLEUROTOMELLA**, Verrill.

Am. Journ. Science, v. 15. 1873.

1. P. PACKARDII, Verrill.

Am. Journ. Science, v. 15. 1873.

Shell thin, fragile, translucent, pale flesh-colored, moderately stout, with an acute, somewhat turreted spire; whorls nine, the apical whorls, for about two and a half turns, nearly smooth, regular, convex, chestnut-colored; below this the whorls are shouldered, strongly convex in the middle, but with a smooth concave band below the suture, corresponding to the posterior notch in the outer lip; the whorls are crossed below the subsutural band by about sixteen strong, prominent, rounded, somewhat oblique ribs, most prominent on the middle of the whorl, but not angulated; on the last whorl these ribs become very oblique below the middle, and follow the curve of the edge of the lip, nearly fading out anteriorly; the surface between the ribs is marked by faint lines of growth, and by fine, unequal, slightly raised revolving lines, which pass over the ribs without interruption. They become more evident on the lower part of the last whorl, and are very faint on the subsutural band, which is more decidedly marked by receding, strongly curved lines of growth. The aperture is rather broad above, elongated below, sub-oval, outer lip very thin, sharp, prominent above, separated from the preceding whorl by a wide and very deep sinus, extending back for about one-fifth of the circumference of the whorl; the anterior border of the lip is incurved near the end and obliquely truncate, forming a short, straight canal. Columella simple, nearly straight, its inner edge towards the end sharp and obliquely excurved.

Length 22, diam. 12 mill.

Bay of Fundy, etc.

Genus **MANGELIA**, Risso.

Moll. Eur. Merid. f. 101. 1826.

1. M. RUBELLA, Kurtz and Stimpson.

Proc. Bost. Soc. Nat. Hist., iv. 115. 1851.

Shell ovately fusiform, longitudinally costate, ribs elevated, acute, spirally finely striate; whorls seven, angulate, the last three-fourths the length of the shell; aperture narrow, one-half the total length, lip thickened, scarcely sinuate; ribs wax-colored, interstices darker, suture porcellanous, last whorl numerously fasciate.

Length 10, diam. 4.5 mill.

Oak Island, North Carolina.

This species I have not seen, and it has not been figured.

2. M. FILIFORMIS, Holmes. Fig. 86.

Post-Pliocene Foss. So. Car. 69, t. 11, f. 9. 1860.

Shell fusiform, costate, transversely striate; costæ smooth, somewhat angulated at the periphery of the middle whorls; the third whorl cancellate; apex smooth; base without ribs; striæ transverse, numerous, slightly waved, filiform.

South Carolina.

In general outline this little shell resembles the young of *Fusus cinereus*, Say, but the smooth ribs, nodulous and angulated periphery of the middle whorls and beautiful filiform striæ, readily distinguish it from that species.

3. M. LABECULA, Gould.

Proc. Bost. Soc. Nat. Hist., viii. 281. 1862.

Shell small, ovate fusiform, waxy, with chestnut markings within and without the lip, thinly spirally striate, with sixteen longitudinal plications which become evanescent on the body-whorl; whorls seven to eight with a subsutural impressed line; aperture small, oval; lip gibbous, thickened without and granulate within. Sinus scarcely perceptible.

Length 7, diam. 3 mill.

(Dredged.)   *Coast of Georgia.*

Family VOLUTIDÆ.

Shell turreted or convolute; aperture notched in front; columella obliquely plaited. No operculum (typically). Animal with a recurved siphon; foot very large, partly hiding the shell: mantle often lobed and reflected over the shell, eyes on the tentacles or near their base.

Synopsis of Genera.

Shell ventricose, thick; spire short, apex mammillated; aperture large, outer lip not thickened, deeply notched in front; columella with several plaits.   VOLUTA, Linnæus.

Shell fusiform, thick; spire elevated, acute; aperture small, notched in front; columella obliquely plaited; operculum very small.

MITRA, Humphreys.

Shell smooth, bright; spire short or cancelled; aperture truncated in front; columella plaited; outer lip of adult with thickened margin.

MARGINELLA, Lamarck.

Genus VOLUTA, Linnæus.

Syst. Nat., edit. x. 729. 1758.

1. V. JUNONIA, Chemnitz. Fig. 87.

Conch. Cab., xi. t. 177, f. 1703, 4. 1799.

Shell fusiformly ovate, spire short, obtuse at the apex; whorls smooth or minutely decussately striated, rather swollen around the upper part; columella strongly four-plaited; aperture rather long; lip simple; cream color, regularly painted with rows of deep red spots.

Georgia, Florida.

This is the most rare and valuable American marine shell; specimens in good condition sell for fifty to one hundred dollars. Its occurrence has been noticed at a number of localities, but it is never plentiful.

Genus MITRA, Humphreys.

Mus. Callon. 1797.

1. M. GRANULOSA, Lamarck. Fig. 88.

Anim. s. Vert., vii. 304. 1822.

Shell oblong, ovate, spire acuminately turreted, sutures rather deep; whorls closely decussated with granular ridges; columella four-plaited, slightly umbilicated; aperture rather short.

North Carolina, W. Indies.

Genus MARGINELLA, Lamarck.

Prodr. 1799.

1. M. GUTTATA, Dillwyn. Fig. 89, *a, b.*

(*Voluta.*) Desc. Cat. 526. 1817

Marginella longivaricosa, Lamarck, Anim. s. Vert., vii. 358. 1822.

Shell somewhat pyriformly oblong, flesh-colored, obscurely two-banded, flaked throughout with opaque white spots; spire small, scarcely exserted, callous; whorls rather gibbous around the upper part; lip broadly thickened, white, distantly spotted with reddish-fawn.

North Carolina to West Indies.

2. M. ROSCIDA, Redfield. Fig. 90.

Proc. Acad. Nat. Sciences Philad., 174. 1860.

Marginella apicina of American authors (not Menke).

Shell rhombic-ovate, polished, grayish-brown, minutely flecked with white ; towards and upon the spire the white spots tend to be confluent in longitudinal lines. Lip white, thickened, obtusely reflected, slightly denticulated within, with three external brown spots. Spire moderate, four whorls visible, the last one shouldered beneath the suture ; columella four-plaited.

Length 14, diam. 8 mill.

North Carolina to Georgia.

Resembles *M. apicina* of West Indies, but distinguished by the white flecked lines; the spire is also more developed and body more angular.

3. M. LACHRYMULA, Gould.

Proc. Bost. Soc. Nat. Hist., viii. 281. 1862.

Shell small, ovate, white, translucent, shining, very finely longitudinally striate ; apex scarcely prominent, vitreous ; aperture narrow, crescentic ; lip obtuse, varicose without and widely arcuate behind, crenulate within ; columella four-plaited.

Length 1.5, diam. 1 mill.

Dredged in 400 fathoms.

Coast of Georgia.

This may be distinguished from one or two other minute species found in the W. Indies by its transparency and its ventricose form.

Family PORCELLANIDÆ.

Shell convolute, enamelled ; spire concealed ; aperture narrow, channelled at each end ; outer lip of adult thickened, inflected. No operculum.

Animal with a broad foot, truncated in front ; mantle expanded on each side, forming lobes, which meet over the back of the shell ; these lobes are usually ornamented with tentacular filaments ; eyes on the middle of the tentacles or near their base ; branchial plume single.

Synopsis of Genera.

Shell convolute, smooth, inner lip crenulated.   PORCELLANA, Rumph.

Shell covered with transverse ribs, inner lip crenulated. TRIVIA, Gray.

Shell smooth, each end produced into a canal ; inner lip smooth, outer lip reflected, thickened externally.   VOLVA, Bolten.

Genus PORCELLANA, Rumphius.
Amboin. Rarit. 113. 1705.

Klein, Tent. Method., 83. 1753.

Cypræa, Linn., Syst. Nat., 1172. 1767.

The Cowries or Cypræas, as they are generally called, are generally medium or large size shells with a handsomely painted enamelled surface, covering and concealing the spire in the adult. This enamel is derived from the lobes of the mantle, which envelop the shell, meeting on the back, where the contact of the margins is indicated by a line of lighter color. In the young shell the spire is prominent, the outer lip thin, and the general appearance is much like that of an *Oliva*. The animal has the power of partially dissolving, and then breaking its shell across the back when it is necessary to provide room for its growth ; it is, in fact, easier to do this than to absorb the immense thickness of the lips previously to adding to the circumference in the ordinary manner of shell-growth.

There are about one hundred and fifty species, inhabiting shallow water near the shore, in warm latitudes, and feeding on zoophytes.

1. P. EXANTHEMA, Linnæus.   Fig. 91.
   Syst. Nat., edit. xii. 1172. 1767.

Shell elongately ovate, rather thick, extremities slightly truncated; back fulvous-brown, ornamented with round white spots; base pale fulvous, teeth chestnut-brown.

North Carolina, West Indies.

A single specimen has been found at Fort Macon, N. C. It will, doubtless, be discovered at other points on the southern coast.

Genus TRIVIA, Gray.
Desc. Cat. Cypr. 1832.

1. T. QUADRIPUNCTATA, Gray.   Fig. 92.
   Zool. Journ. iii., 368. 1827.

C. rotunda, Kiener, Coq. Viv., 141, t. 53, f. 2.

Shell rotundately ovate, extremities obtuse, transversely very finely ribbed, with a linear dorsal groove; light rose color, back ornamented with four conspicuous red dots, two on each side of a dorsal groove, alternating one with the other.

North Carolina to West Indies.

Genus **VOLVA**, Bolten.

Museum. 1798.

1. V. UNIPLICATA, Sowerby. Fig. 93.

(*Ovulum.*) Proc. Zool. Soc., 135. 1848.

Shell narrowly oblong, rather thin, whitish or orange-rose, extremities rather produced, blunt, back transversely minutely striated; lip moderately thickened, flexuous, widely sinuated at the lower part; columella plicate above.

South Carolina, southwards.

ORDER I. **PROSOBRANCHIATA**.

Section B. HOLOSTOMATA. Sea-snails.

Shell spiral or limpet-shaped; rarely tubular or multivalve; margin of the aperture generally entire; operculum horny or shelly, usually spiral. Animal with a short non-retractile muzzle; respiratory siphon wanting, or formed by a lobe developed from the neck; gills pectinated or plume-like, placed obliquely across the back, or attached to the right side of the neck; neck and sides frequently ornamented with lappets and tentacular filaments. Mostly phytophagous. Aperture channelled in front, with a less distinct posterior canal.

I. *Aperture entire. Shell globular or turbiniform.*

Family I. NATICIDÆ. Shell globular, few-whorled; spire small, obtuse; aperture semilunar; lip acute; pillar often callous.

II. *Shell elongated.*

Family II. PYRAMIDELLIDÆ. Shell spiral, turreted; nucleus minute, sinistral; aperture small; columella sometimes with one or more prominent plaits; operculum horny, imbricated; nucleus internal.

Family III. CERITHIADÆ. Shell spiral, elongated, many-whorled, frequently varicose; lip generally expanded in the adult; operculum horny and spiral.

Family V. TURRITELLIDÆ. Shell tubular or spiral, upper part partitioned off; aperture simple; operculum horny, many whorled.

Family VI. LITTORINIDÆ. Shell spiral, furbinated, or depressed, never pearly; aperture rounded; peristome entire; operculum horny, pauci-spiral.

Family IX. TURBINIDÆ. Shell spiral, turbinated, or pyramidal, nacreous inside; operculum calcareous and pauci-spiral, or horny and multispiral.

III. *Shell ear or limpet shaped, scarcely spiral; aperture basal, very large.*

Family X. JANTHINIDÆ. Shell spiral, turbiniform; fragile; aperture large; outer lip notched in the middle; no operculum.

Family XI. FISSURELLIDÆ. Shell conical, limpet-shaped; apex recurved; nucleus spiral, often disappearing in the course of growth; anterior margin notched or apex perforated; muscular impression horse-shoe shaped, open in front.

Family XII. CALYPTRÆIDÆ. Shell limpet-like, with the apex more or less spiral; interior simple or divided by a shelly process, variously shaped, to which the adductor muscles are attached.

Family XIII. PATELLIDÆ. Shell conical, limpet-shaped, the apex entire and turned forwards; muscular impression horse-shoe shaped, open in front.

Family XV. CHITONIDÆ. Shell oval, *composed of eight transverse imbricating plates*, lodged in a coriaceous mantle, which forms an expanded margin round the body.

IV. *Shell tubular, symmetrical, open at both ends; not spiral.*

Family XIV. DENTALIADÆ. (The tooth shells.)

Family I. NATICIDÆ.

The Natices, although possessing shells with entire apertures, are carnivorous animals, feeding principally on small bivalves, boring small round holes through their shells and rasping out the meat.

Synopsis of Genera.

Shell sub-globose; spire rather elevated; aperture semilunar; columella adherent to and spirally produced into the umbilicus; apex more or less dilated and truncate; more rarely convex or rounded; operculum horny, with a calcareous outer layer.   NATICA, Adanson.

Shell oval, subglobose; spire rather elevated; aperture semilunar; inner lip thin or with a moderate callus; umbilicus wide, pervious, not funiculate. Operculum simple, cartilaginous.   LUNATIA, Gray.

Shell orbicular, depressed; spire flattened; aperture wide, semilunar; inner lip straight, callous; columella adherent to and spirally contorted into the umbilicus; apex more or less dilated and truncate. Operculum simple, cartilaginous.   NEVERITA, Risso.

Shell ovate or subovate, solid, smooth, usually without epidermis; spire small, acute, whorls simple; aperture semicircular; inner lip oblique, thickened, callous; umbilicus funiculate; aperture adherent to and spirally contorted into the umbilicus, the apex more or less dilated, convex, and rounded. Operculum large, horny, simple. MAMMA, Klein.

Shell ventricose, imperforate; spire with the apex acute; whorls smooth, without epidermis; aperture very wide; inner lip with a large smooth callus covering part of the body-whorl and concealing the umbilicus.
BULBUS, Bronn.

Shell longitudinally oval, thin, smooth, white, not umbilicated, covered with a light-brown epidermis; spire elevated, suture canaliculated; aperture oval, produced in front. Operculum pauci-spiral, horny, thin.
AMAUROPSIS, Mörch.

Shell oval, flattened, ear-shaped, striated; spire minute, depressed; aperture very wide, oblique; umbilicus none; inner lip curved posteriorly and spread thinly over the body-whorl. Operculum very small and rudimentary. SIGARETUS, Lamarck.

Shell thin, small, ear-shaped, pellucid, fragile; spire very small; aperture large, patulous; inner lip receding. No operculum.
MARSENINA, Gray.

Shell thin, small, subglobose, composed of two rapidly enlarging volutions; aperture large, subovate; lip thin, not joined behind; usually covered with a velvety epidermis. No operculum. VELUTINA, Blainville.

Genus NATICA, Adanson.
Hist. Nat. Senégal, 172. 1757.

1. N. AFFINIS, Gmelin. Fig. 94.
(*Nerita.*) Syst. Nat., 3675. 1790.
N. clausa, Brod. et Sowb. Zool. Journ., iv. 360. 1829.
N. consolidata, Couthouy, Bost. Journ. Nat. Hist., ii. 89, t. 3, f. 14. 1838.

Shell subglobose; whorls four or five, subconvex, partially flattened or even concave near the sutures; spire very short, obtuse; suture distinct; aperture oval, widest above; lip sharp, thickened, and rounded towards the umbilicus; callus depressed, enlarged at the upper angle, and in mature specimens quite concealing the umbilicus; epidermis thin, greenish-brown. Operculum and throat milk-white.

Length 12.5, diameter 14 mill.

Obtained from stomachs of fishes.

Cape Cod, Mass., northwards (Eur.).

2. N. PUSILLA, Say. Fig. 95.
Jour. Philada. Acad. Nat. Sciences, ii. 257. 1822.

Shell thin, suboval, cinereous, or rufous, with sometimes one or two obsolete, dilated, revolving bands; columella callous; callus pressed laterally into the umbilicus, whitish; umbilicus nearly closed and consisting only of an arcuated, linear, vertical, aperture.

Length about 6 mill.

Buzzard's Bay. Dredged in three to eight fathoms. (Stimpson.)

Massachusetts to N. Carolina.

Genus LUNATIA, Gray.

Zool. Proc. 1847.

1. L. HEROS, Say. Fig. 96, 97 (*triseriata*), 98 (*Nidus*), 99 (*Animal*).

(*Natica.*) Journ. Philada. Acad., ii. 248. 1822.

N. catenoides, S. Wood, 1848 (teste Jeffreys.)

Var. N. triseriata Say. Journ. Phila. Acad., v. 209.

Shell large, thick, globular-ovate; whorls five, convex; spire considerably elevated. Aperture oval; the callus reflected over a small portion of the large, patulous, and coarsely wrinkled umbilicus. Epidermis thin and yellowish; beneath this, ashen gray. Aperture dark reddish-brown occasionally tinged with yellowish.

Occasionally grows to the length of five inches. The young shell is sometimes marked by three revolving series of parallel oblong brown spots, twelve or fourteen in each row (*N. triseriata*). These markings are lost on the subsequent whorls.

The nidus is composed of sand agglutinated into a bowl-shaped mass, open at the top. The eggs are attached to the interior surface.

This may be the *N. ampullaria* of Lamarck, as surmised by some of our conchologists, but the description does not sufficiently characterize it.

New England, southward to New Jersey.

2. L. GRŒNLANDICA, Möller. Fig. 100.

(*Natica.*) Faun. Grœnl., 7. 1842.

Natica pusilla, Gould (not Say), Invert. Mass., edit. i. 237, f. 166. 1841.

Shell suboval, smooth, glossy or with faint incremental and revolving lines; whorls four, regularly rounded; spire moderately elevated, obtuse; suture distinct and deep; lip sharp, acute; callus pressed laterally into the umbilicus, leaving a narrow curved linear opening. Epidermis ash-colored, beneath bluish-white.

Length 12.5, diam. 11 mill.

Massachusetts northwards.

Genus NEVERITA, Risso.
Moll. Eur. Merid., iv. 149. 1826.

1. N. DUPLICATA, Say. Fig. 101.

(*Natica.*) Journ. Phila. Acad. Nat. Sci., ii. 247. 1822.

Shell solid, subglobose; whorls five, somewhat convexly flattened above; aperture oval, oblique; umbilicus irregular, with a deep furrow, and almost entirely covered by a thick callus. Color ashen-gray, within deep chestnut-brown; callus of the same color.

Length 25–50, diam. 27–55 mill.

New England to Georgia.

Genus MAMMA, Klein.
Ostracol. 21. 1753.

1. M. IMMACULATA, Totten. Fig. 102.

(*Natica.*) Amer. Journ. Sci., xxviii. 357 f. 6. 1835.

Shell small, solid, longitudinally suboval. Whorl about five, the upper ones very slightly convex; apex short and pointed, suture not impressed; the body-whorl convex and elongated beneath. Aperture narrow, regularly and somewhat acutely curved at the base. Umbilicus rounded and deep, scarcely modified by the callus, which is not very copious, but forms a deposit under the upper part of the lip, and causes a white spiral line to appear externally just below the suture. Epidermis thin, greenish-yellow, beneath milk white.

Length 7, diam. 5.5 mill.

Maine to New York.

Genus BULBUS, Brown.
Proc. Geol. Soc., iii. 119. 1839.

1. B. SMITHII, Brown. Fig. 103.

Natica flava, Gould, Am. Journ. Science, xxxviii. 196. 1840.
Natica aperta, Lovén, Index Moll. Scand. 1846.

Shell globular, inflated, thin and light; whorls four, rounded, slightly compressed above near the suture, with very minute incremental and revolving striæ; spire little elevated; pillar-lip with a curve in its middle; the callus contracting and obliterating the umbilicus, which is deeply indented; epidermis light yellowish, white underneath; callus ivory white.

Maine northwards (Eur.).

Genus AMAUROPSIS, Mörch.

1. A. ISLANDICA, Gmelin. Fig. 104.
(*Nerita.*) Syst. Nat. 3675. 1790.
Natica helicoides, Johnston, Trans. Berw. Club, i. 69, 266. 1835.
Natica canaliculata, Gould, Am. Journ. Sci., xxxviii. 197. 1840.

Shell subglobose, rather ponderous, dingy-white, nearly smooth and somewhat glossy, covered with a dark gamboge-colored epidermis; whorls four, the upper portion of each turning before it joins the preceding whorl, so as to form a broad, shallow canal at the suture, and giving the spire a turreted appearance; aperture about two-thirds the length of the shell, nearly semicircular; lip sharp, a little spreading in front, the inner margin nearly a straight line, and overspread with a thick callus; interior white; umbilical opening a mere slit, generally none.

Length 27.5, diam. 18 mill.

Resembles in shape *Paludina ponderosa.*

Massachusetts northwards (Eur.).

Genus SIGARETUS, Lamarck.

Prodr. 77. 1799.

H. and A. Adams, and others, use the generic name *Catinus*, Klein, in preference to that given by Lamarck, but Klein's genus is named "*Catinus lactis*," and is, therefore, not admissible.

1. C. PERSPECTIVUS, Say. Fig. 105.
Am. Conch., iii. t. 25. 1831.

Shell ovate, elongate, depressed. Surface with numerous impressed, transverse, slightly undulated lines, crossed by revolving striæ which become obsolete beneath. Aperture more than three-fourths the entire area of the shell. Whorls three; spire depressed, smooth; suture distinct, but not deeply impressed. Milk-white, sometimes tinged with brown; within smooth and polished, and faintly iridescent.

Length 37, of aperture 24 mill.

New York, southwards.

2. C. MACULATUS, Say. Fig. 106.
Am. Conch., iii. t. 25. 1831.

Shell with numerous transverse hardly undulated impressed lines and longitudinal wrinkles; spire scarcely prominent, slightly convex; whorls about three; suture a simple impressed line.

Color whitish, with two bands of pale rufous spots and a rufous band near the suture.

Southern Coast.

A doubtful species; has not been recently detected.

Genus MARSENINA, Gray.

1. M. PERSPICUA, Linnæus. Fig. 107.
   (*Helix.*) Syst. Nat., edit. xii. 1260. 1767.
Helix haliotoidea, Linn. Syst. Nat., edit. xii. 1250. 1757.
Sigaretus haliotoideus, Lamarck, Anim. s. Vert., vi. 208. 1822.
Oxynoe glabra, Couthouy, Bost. Journ. Nat. Hist., ii. 90, t. 3, f. 10. 1838.

Shall small, obliquely ovate, pellucid, white, compressed, smooth; aperture very large; whorls two.

Length 12.5, diam. 10 mill.

Coast of Massachusetts; Northern Europe.

Genus VELUTINA, Blainville.

Malacol. 468. 1825.

1. V. LÆVIGATA, Linnæus. Fig. 108.
   (*Helix.*) Syst. Nat., edit. xii. 1250. 1767.
Helix haliotoidea, Fabricius, Fauna Grœnl. No. 387. 1780.
Bulla velutina, Müller, Zool. Dan. iii. t. 101, f. 1-4. 1789.
Velutina capuloidea, Blainville, Malacol. t. 42, f. 4. 1825.
Velutina rupicola, Conrad. Journ. Phila. Acad. Nat. Sci., vi. 266. 1831.

Shell small oval, very thin and fragile. Whorls three; the body-whorl with faint concentric striæ; spire slightly raised, smooth at the apex. Aperture regularly oval. Epidermis, when not abraded, thick and raised, more conspicuous on the concentric lines. Dusky brown with numerous revolving raised rufous lines.

Length 7.5, diam. 6.25 mill.

Cape Cod, Massachusetts, northwards (Eur.).

This shell is extremely fragile, seeming to consist principally of epidermis. European specimens attain larger dimensions and are more solid.

2. V. ZONATA, Gould. Fig. 109.
   Invert. Mass., edit. i. 242, f. 160. 1827.

Shell small, ovate, moderately thin; spire not raised. Whorls three; the two upper ones scarcely distinct; suture deeply impressed, striated with revolving lines and superficial concentric furrows. Aperture regularly oval; lip expanded, exceedingly thin and fragile; pillar lip flattened, and with a small superficial fold. Epidermis whitish or reddish-brown, with numerous bands of brown; pillar white.

Length 10, diam. 12.5 mill.

Generally covered with a calcareous incrustation; underneath zoned or frequently entirely white or flesh color.

Coralline, on stones, dredged in 14 to 40 fathoms.

Massachusetts northwards (Eur.).

Family PYRAMIDELLIDÆ.

The animal, in this family, is provided with broad, ear-shaped tentacles which are often connate, with eyes behind them at their bases; proboscis retractile; foot truncated in front; the tongue unarmed.

Synopsis of Genera.

Shell subulate, turreted, many-whorled, smooth; spire pointed, nucleus sinistral; aperture semi-oval, entire, rounded anteriorly; columella straight, plicated; outer lip acute. OBELISCUS, Humphrey.

Shell slender, elongated, many-whorled, longitudinally ribbed; apex of spire with a persistent, embryonic, sinistral nucleus; aperture oblong or subquadrate, peristome incomplete; columella straight, simple, edentulate and without a plait. TURBONILLA, Risso.

Shell turreted, subulate or ovate, smooth or transversely striated; apex of spire sinistral; aperture ovate, peristome not continuous; columellar lip with a single tooth-like fold. ODOSTOMIA, Fleming.

Shell turreted, many-whorled, whorls smooth or spirally striated; aperture oval or rounded; inner lip simple, without plaits or teeth, base often perforated. ACLIS, Lovén.

Shell elongated, imperforate, many-whorled, transversely striated; aperture small, ovate, entire in front; columellar lip smooth or obscurely plaited. MENESTHO, Möller.

Shell elongated, white, smooth, polished; spire produced, many-whorled, frequently with an interrupted varix on one side, apex acute; aperture oval, pointed behind; inner lip reflected over the pillar; axis imperforated; outer lip thickened internally. EULIMA, Risso.

Shell subulate or subglobose, thin, pellucid, smooth, polished, many-whorled; apex of spire produced and styliform, with a sinistral nucleus; aperture sub-ovate, pointed behind, rounded and entire in front; inner lip smooth, arcuated; outer lip thin, simple. STYLIFER, Broderip.

Genus **OBELISCUS**, Humphrey.

Mus. Calonn. 1797.

1. O. CRENULATUS, Holmes.   Fig. 110.

Post Pliocene Foss. So. Car. 88, t. 13, f. 14.   1860.

Shell subulate, smooth, angularly channelled at the suture, which
is crenulated on the lower whorls only; columella with three folds,
the superior one large and acute, the others small; outer lip with
four small teeth internally; whorls fourteen to sixteen; with two
indistinct opaque revolving bands, which are sometimes obsolete.

North and South Carolina.

Genus **TURBONILLA**, Risso.

Hist. Nat. Eur. Merid., iv. 224.   1826.

Chemnitzia, Orb. Webb and Berthol. Hist. Nat. Iles Canaries.   1830.

1. T. INTERRUPTA, Totten.   Fig. 111.

(*Turritella*) Am. Journ. Science, xxviii. 352, f. 7.   1834.

Shell small and slender, whorls ten, almost flat, on which are
twenty to thirty transverse obtuse ribs, crossed by about fourteen
subequal revolving lines interrupted by the ribs; these are
arranged in pairs, so closely applied as often to be confounded in
one; below the middle of the body-whorl, the ribs become obsolete,
and the revolving lines are uninterrupted; a slight shoulder on
each whorl causes the sutures to be quite distinct. Aperture
ovate, sharply angular above; inner lip slightly everted. Whitish-
brown or amber-colored.

Length 6.25, diam. 2.4 mill.

Mass. to North Carolina.

2. T. NIVEA, Stimpson.   Fig. 112.

Proc. Bost. Soc. Nat. Hist., iv. 114.   1851.

Shell aciculated, sub-cylindrical, white shining; whorls eleven,
flattened, longitudinally plicate; folds straight, interstices per-
fectly smooth.

Length 7, diam. 1 mill.

Animal white; head short; tentacles triangular, very broad,
with the eyes at nearly the middle of their bases; foot elongated,
with an arcuated indentation at its anterior terminus.

This species differs from *T. interrupta* in being more slender, in
wanting revolving lines, and also totally in its station, the deeper
parts of the Coralline Zone. Taken in forty fathoms, on a muddy
and gravelly bottom.

Grand Manan Island, Bay of Fundy.

3. T. ELEGANS, Verrill. Fig. 113.
Am. Journ. Science, 3d ser. iii. 282, t. 6. f. 4. 1872.

Shell light yellowish, elongated, moderately slender, acute. Whorls ten or more, well rounded, not distinctly flattened; suture rather deeply impressed; surface somewhat lustrous, with numerous rounded vertical costæ, narrower than the concave interspaces, fading out below the middle of the last whorl; and with numerous fine revolving grooves, interrupted on the costæ; on the upper whorls there are about five; and on the lower half of the last whorl usually five or six distinct and continuous ones. Aperture broad oval, anteriorly rounded and slightly effuse; outer lip thin, sharp; columella nearly straight at base within, slightly revolute, outwardly, regularly curved anteriorly where it joins the outer lip, and not forming an angle with it. The epidermis is thin, light yellow, sometimes with a darker, yellowish revolving band on the middle of the last whorls, and also with the revolving striæ darker.

Length 5, diam. 1.7 mill.

Less slender than *T. interrupta* with more rounded whorls. (Dredged in 8–10 fathoms.)

Vineyard Sound.

Doubtful and Undetermined Species.

4. T. TEXTILIS, Kürtz.
Cat. Mar. Shells, 8. 1860.

Whorls six or seven, shouldered, thick-set with prominent, smooth, longitudinal ribs, the interspaces crossed by impressed revolving lines. White, waxy or chalky.

Length 3, diam. 1.15 mill.

Fort Johnson, S. C.

5. T. SPIRATA, Kürtz and Stimpson.
(*Chemnitzia.*) Bost. Proc., iv. 115. 1851.

Shell ovate-conical, umbilicate, white, shining; with minute transverse striæ; whorls six, flat, angulate in front; suture profound; aperture small, ovate; columella edentulous.

Length 2.5, diam. 8 mill.

North Carolina.

6. T. CURTINA, Gould.
(*Chrysallida.*) Bost. Proc., viii. 280. 1862.

Shell minute, ovately turreted, whitish; whorls six to seven, convex, cancellate by four series of granules; last whorl about

5 ·

half the total length; aperture narrow, produced in front; columella with a minute posterior plica.

Length 5, diam. 2 mill.

South Carolina.

It has the aperture, and in general the sculpture of *Chemnitzia*, with the columellar fold of *Odostomia*.

7. T. SUTURALIS, Gould.

(*Dunkeria.*) Bost. Proc., viii. 280. 1862.

Shell minute, ivory white, turreted; whorls seven to eight, convex, with ten to twelve longitudinal plications (evanescent above) and thin revolving striæ, of which a subsutural one is well impressed; last whorl less than half the total length; aperture lunate, small.

Length 3, diam. 1 mill.

Fort Johnson, Charleston, S. C.

Genus ODOSTOMIA, Fleming.

Brit. Anim. 297, 310. 1828.

This genus consists of small, usually white, smooth, solid and enamelled shells, with the inner lip always toothed; they range from low-water to forty fathoms.

1. O. PRODUCTA, Adams. Fig. 114.

(*Jaminea.*) Bost. Journ. Nat. Hist., iii. 322, t. 3, f. 8. 1840.

Shell small, conic-cylindrical; whorls eight, nearly flat; epidermis light-brown; columella flexuous.

Length 5 millimetres.

More slender than *O. fusca*, with one or two more whorls, which are less convex and with no approach to an umbilicus.

Massachusetts.

2. O. FUSCA, Adams. Fig. 115.

(*Pyramis.*) Bost. Journ. Nat. Hist., ii. 282, t. 4, f. 9. 1839.

Shell small, subelongate, conical; spire truncate, obtuse; whorls six, convex; suture strongly impressed, and with a revolving line below it, causing it to appear double. Aperture broadly ovate, acutely angular above, dilated in the middle. Fold on the pillar-lip far within, occasionally double, and in some cases obsolete; an umbilical indentation about the middle of the left lip. Epidermis shining brown.

Length 6 mill.

New England; New York.

Closely allied to *O. bisuturalis*, but is shorter, and spire less acute ; whorls flatter and color darker.

3. O. DEALBATA, Stimpson. Fig. 116.
(*Chemnitzia.*) Proc. Bost. Soc. Nat. Hist., iv. 114. 1851.

Shell ovate-conic, white, smooth, pellucid ; whorls six, rather convex ; aperture ovate, hardly effuse ; furnished with a small inconspicuous fold.

Length 4, diam. 1.5 mill.

It is broader than *O. bisuturalis*, but has not so sharp an apex, and wants the revolving line.

Dredged in three fathoms, on a shelly bottom.

Boston Harbor.

4. O. MODESTA, Stimpson. Fig. 117.
(*Chemnitzia.*) Proc. Bost. Soc. Nat. Hist., iv. 16. 1851.

Shell small, conic, white, smooth ; whorls four, flattened, the last medially sub-angulated ; suture impressed ; aperture uniplicate, sub-rhomboid.

Length 3.5, diam. 1.5 mill.

This species is more angular than *O. bisuturalis*, and has no revolving line just below the suture as in that shell. It is very like the British *O. unidentata.* It inhabits the Coralline Zone.

• St. George's Banks.

5. O. BISUTURALIS, Say. Fig. 118.
(*Turritella.*) Journ. Philad. Acad., ii. 244. 1821.
Chemnitzia bisuturalis, Stimpson, Shells of N. England, 42. 1851.
Jaminea exigua, Couthouy, Bost. Journ. Nat. Hist., ii. 92, t. 2, f. 7. 1838.
Odostomia exigua, Gould, Invert. Mass., 1st Edit. 272, f. 77. 1841.
Rissoa rupestris, Forbes, Ann. Nat. Hist., ii. 107, t. 2, f. 13.

Shell ovate-conical, apex rather obtuse, smooth, light-green, epidermis brownish ; whorls five or six, flattened, with an impressed line revolving below the suture, giving the appearance of a double suture ; columella with a transverse fold.

Length 5, diam. 2 mill.

New England.

6. O. TRIFIDA, Totten. Fig. 119.
(*Actæon.*) Am. Journ. Science, xxvi. 368, t. 1, f. 4, *a, b.* 1834.

Shell small, elevated, pointed, smooth and glossy ; whorls eight, flat, with about six impressed revolving lines ; the one above and the two next below the suture wider and more distinct ; ten or twelve very minute lines at the base of the body-whorl. Spire gradually tapering to an acute apex. Aperture elongated, about one-third the length of the shell, acutely angular above, produced and rounded below. Outer lip sharp and thin, entire; pillar lip with a single sharp, oblique fold; opercule horny; ivory or soiled white.

Length 5, diam. 2 mill.

Mr. Gwyn Jeffreys believes this to be a variety of *O. impressa,* Say.

New England; New York.

7. O. SEMINUDA, Adams.   Fig. 120.

(*Juminea.*)   Bost. Journ. Nat. Hist., ii. 280, t. 4, f. 13.   1839.

Shell small, acute, conic. Whorls seven, convex; upper whorls and half of the body-whorl longitudinally rugose, crossed by three equidistant revolving lines, presenting a granulated appearance. At the base of the lower whorl are four revolving lines, beginning on the middle, where the folds abruptly terminate. Suture distinct, divided by an indistinct spiral ridge. Aperture oval; the outer lip very thin, and scalloped by the revolving lines; the pillar lip with an inconspicuous fold. Glossy white, translucent.

Length 3.75, diam. 1.75 mill.

New England, southwards.

8. O. IMPRESSA, Say.   Fig. 121.

(*Turritella.*)   Journ. Philad. Acad., ii. 244.   1822.

Odostomia insculpta, De Kay, Moll. N. Y. 115, t. 31, f. 297.   1843.

Shell elevated, thick, opaque, regularly tapering to the apex. Whorls seven, flat, with a deeply impressed suture; body-whorl with ten deeply sculptured closely approximated revolving striæ on the lower half, and five distant revolving lines on the upper half; about four on the next whorl, and gradually diminishing in number above. Aperture ovate, acute above, effuse beneath. Lip simple; fold on the pillar-lip near the middle, distinct under the lens, and deepening within. Soiled white, the sculptured lines rufous.

Length 5, diam. 2 mill.

New England, southwards.

Genus **ACLIS**, Lovén.

Ind. Moll. Scand. 16.   1846.

The animal has a linguiform foot, much produced anteriorly,

tentacles cylindrical, much swollen at their tips. The species are few in number and small, though exceedingly elegant.

1. A. POLITA, Verrill. Fig. 122.

Am. Journ. Science, 3d series, iii. 282, t. 6, f. 5. 1872.

Shell white, elongated, regularly tapering, slender, acute. Whorls thirteen or more, convex, rounded, scarcely flattened; surface smooth, polished, shining, with faint or scarcely distinct striæ of growth. Aperture broad oval; outer lip sharp, slightly effuse; columella slightly curved, without a fold.

Length 8, diam. 2 mill.

Dredged in twenty fathoms.

Eastport Harbor, Mᵉ.

Genus MENESTHO, Möller.

Index Moll. Scand. 1842.

1. M. STRIATULA, Couthouy. Fig. 123.

(*Pyramis.*) Bost. Journ. Nat. Hist., ii. 101, t. 1, f. 6. 1838.

Menestho albula, Möller, Moll. Scand. 1842.

Menestho striata, Chenu. Man. Conchyl., i. 229, f. 1311.

Shell smooth, subulate, imperforate, usually polished. Whorls seven to nine, nearly flat, marked by twelve to fifteen minute regular revolving striæ, diminishing in number to the apex; suture linear and rather deeply impressed. Aperture ovate acute, angular above; base very slightly effuse; outer lip sharp, smooth, without any sinus or groove at its junction with the body-whorl; pillar lip arched regularly throughout. Pale bluish-white, milk-white within.

Length 15, diam. 5 mill.

New England.

Genus EULIMA, Risso.

Hist. Nat. Eur. Merid., iv. 123. 1826.

The Eulimæ crawl with the foot greatly in advance of the head, which is usually concealed beneath the margin of the shell; many of them have distorted shells, the upper whorls being often curved or inclined to one side.

1. E. OLEACEA, Kurtz and Stimpson. Fig. 124.

Proc. Bost. Soc. Nat. Hist., iv. 115. 1851.

Shel small, subulate, solid, very shining, white, marked with light-brown transverse bands; whorls twelve, flattened, closely coiled; suture inconspicuous; aperture small, ovate.

Length 6.25, diam. 1.5 mill.

The animal is white, hyaline; tentacles almost joining each other at their bases, where, on the external sides, are the eyes, which may be seen through the shell, when, as is usually the case, the head does not project beyond it. Foot short, broad, slightly produced at the anterior angles; the lobe above projecting a little beyond it.

Dredged in eight fathoms, on a muddy bottom.

New England, North Carolina.

2. E. CONOIDEA, Kurtz and Stimpson.

Proc. Bost. Soc. Nat. Hist., iv. 115. 1851.

Shell conic-lanceolate, white, very shining; with thirteen flat whorls, the last subangulate; aperture rhomboidal.

Length 9, diam. 2.5 mill.

Dredged on muddy bottoms, shallow water.

North and South Carolina.

The above is a copy of the original description. The species has not been figured, and I have never seen a specimen.

Genus STYLIFER, Broderip.

1. S. STIMPSONII, Verrill.

Am. Journ. Science, 3d ser. iii. 283. 1872.

Shell white, short, swollen, broad-oval; spire short, rapidly enlarging. Whorls four or five, the last one forming a large part of the shell; convex, rounded, with the suture impressed, surface smooth, or with very faint striæ of growth; a slightly impressed revolving line just below the suture. Aperture large and broad.

Length 3.75, diam. 3 mill.

Parasitic on Eurycchinus Dröbachiensis, V, in 32 fathoms.

New Jersey, New England, northwards.

Family III. CERITHIADÆ.

Shell spiral, elongated, many-whorled, frequently varicose; aperture channelled in front, with a less distinct posterior canal; lip generally expanded in the adult; operculum horny, pauci-spiral.

Animal with a short muzzle, not retractile; tentacles distant, slender; eyes on short pedicels, connate with the tentacles; mantle margin with a rudimentary siphonal fold.

Synopsis of Genera.

Shell turreted, many-whorled, generally reticulate or granular with indistinct varices ; canal produced in front and slightly recurved ; columella thickened, with a callosity at the hind part. CERITHIUM, Adanson.

Aperture with a slight canal in front, not produced or recurved ; inner lip acute, not reflexed or expanded. BITTIUM, Leach.

Shell sinistral ; aperture round, produced anteriorly into a closed, tubular canal, sometimes with a posterior, closed canal. TRIPHORIS, Deshayes.

Genus CERITHIUM, Adanson.

Hist Nat. Senegal. 153. 1757.

1. C. FERRUGINEUM, Say. Fig. 126.
  Am. Conch., t. 49, f. 3. 1832.

Shell oblong-conical, with longitudinal ribs rendered nodulous by revolving elevated striæ ; ribs about twenty on the body-whorl, almost interrupted by the interstices of the striæ; striæ about seven with intermediate smaller ones on the body-whorl, and but three on the second whorl ; volutions seven, suture inconspicuous ; aperture oblique, oval, whitish within ; labrum slightly thickened on the exterior margin and with obsolete impressed lines on the inner side corresponding with the exterior striæ ; color ferruginous.

North Carolina to Florida.

Genus BITTIUM, Leach.

Gray, Zool. Proc. London. 1847.

This group includes several northern forms, smaller than the typical Cerites, and resembling dextral Triphores; they range from low water to eighty fathoms.

1. B. NIGRUM, Totten. Fig. 127.
  (*Pasithea*) test. juv. Am. Journ. Sci., xxvi. 369, t. 1, f. 7. 1834.
Cerithium reticulatum, Totten, ibid. xxviii. 352, f. 8. 1835.
Cerithium Sayi, Menke, Gould, Invert. Mass. 1st Edit. 278, f. 183. 1841.

Shell small, acute, conic, thin. Whorls from six to eight with a distinct shoulder formed by a series of granules. Surface granular from the crossing of slightly elevated folds with elevated spiral lines ; about twenty of these ribs, which disappear on the lower half of the body-whorl leaving there only about six slightly elevated revolving lines. Suture deeply impressed. Aperture about a fourth of the length of the shell, elongate, subovate, acutely angular above, widely rounded below, slightly effuse. Lip sharp, modified by the revolving lines; the canal, an oblique fissure. Operculum horny, ovate, concave externally, multispiral. Bluish to reddish-black.

Length 5, diam. 2.5 mill.

Cape Cod, Mass., to North Carolina.

2. B. GREENII, Adams. Fig. 128.

(*Cerithium*) Bost. Journ. Nat. Hist., ii. 287, t. 4, f. 12. 1839.

? *Cerithiopsis tubercularis*, Mont. (sp.) 1803, teste Jeffreys.

Shell very small, cylindrical; beneath deeply rugose, with longitudinal ridges and revolving lines. Canal very deep and short, slightly curved. Aperture one-eighth the length of the shell, nearly circular. Reddish-brown.

Length 5, diam. 1.25 mill.

Mass. to North Carolina (Eur.).

Genus TRIPHORIS, Deshayes.

Animal. The tentacles are clavate at the tips, united at their bases by a sinuated veil.

The species are very numerous, inhabiting all seas, but principally tropical. The sculpture of the whorls is very varied, beautiful, and constant.

1. T. NIGROCINCTUS, Adams.   Fig. 129.

(*Cerithium*) Bost. Journ. Nat. Hist., ii. 286, t. 4, f. 11. 1839.

Shell conic-cylindrical, with three revolving series of granules. Whorls reversed, suture broad, carinate; aperture small, subelliptical, ending in a twisted canal about one-third as long as the aperture. Reddish-black; columella black; a black spiral belt in faded shells.

Length 7.5, diam. 1.75.

Massachusetts to North Carolina.

Family V. TURRITELLIDÆ.

Synopsis of Genera.

Shell elongated, many-whorled, spirally striated; aperture rounded, margin thin; operculum horny, many-whorled. TURRITELLA, Lamarck.

Shell tubular, attached; sometimes regularly spiral when young; always irregular in its adult growth; tube repeatedly partitioned off; aperture round; operculum circular, concave externally. VERMETUS, Adanson.

Shell at first discoidal, becoming decollated when adult; tubular, cylindrical, arched; aperture round, entire; apex closed by a mammillated septum; operculum horny, many-whorled. Cæcum, Fleming.

Shell mostly pure white and lustrous; turreted; many-whorled; whorls round, sometimes separate, ornamented with numerous transverse ribs; aperture round; peristome continuous; operculum horny, few-whorled.
SCALARIA, Lamarck.

Genus TURRITELLA, Lamarck.

Prodr. 74. 1801.

The animal in this genus has long, subulate tentacles; eyes slightly prominent; foot truncated in front, rounded behind, grooved beneath; branchial plume very long; lingual ribbon minute, denticulated. Carnivorous?

The species range from low water mark to 100 fathoms; their geographical distribution is extensive, embracing most tropical and temperate seas.

1. T. EROSA, Couthouy. Fig. 130.
  Bost. Journ. Nat. Hist., ii. 103, t. 3, f. 1. 1838.

T. polaris, Möller, Ind. Moll. Scand. 1842.

Shell turreted, elongated; whorls nine to eleven, rather flat, smooth, sloping towards the suture; from three to five abruptly revolving grooves, most prominent and numerous on the lower whorls; striæ of growth wrinkling the shell longitudinally; apex often eroded; aperture circular; lip thin and impressed by the termination of the costæ; columella with a slight callus and angular base; reddish-brown.

Length 12.5, diam. 3.75 mill.

Massachusetts, northwards (Eur.).

2. T. RETICULATA, Mighels and Adams. Fig. 131.
  Bost. Journ. Nat. Hist., iv. 50, t. 4, f. 19. 1842.

T. lactea, Möller, Ind. Moll. Scand. 1842.

Shell turreted, slender, grayish-white; whorls eleven or twelve, convex, with irregular longitudinal folds, and three to five revolving striæ, making the surface appear reticulated; aperture small, orbicular; labrum thin; operculum horny.

Length 17, diam. 5 mill.

Allied to *T. erosa,* but easily recognized by the longitudinal ribs, and by its more slender form.

Gulf of St. Lawrence.

3. T. COSTULATA, Mighels and Adams.   Fig. 132.
  Bost. Journ. Nat. Hist., iv. 50, t. 4, f. 20.   1842.

Shell whitish, translucent; whorls nine or ten, nearly flat or very slightly convex; suture well impressed; last two whorls nearly smooth, the others longitudinally plicate, with microscopic transverse striæ; last whorl subcarinate; aperture rather less than one-fourth the length of the shell, subovate, produced anteriorly.
  Length 17, diam. 6. mill.

Casco Bay.

4. T. ACICULA, Stimpson.   Fig. 133.
  Bost. Proc., iv. 15.   1851.

Shell small, turreted, subulate, white, thin; whorls ten, very convex, longitudinally striate and transversely ribbed; aperture rounded, effuse anteriorly; peristome acute.
  Length 5, diam. 1.5 mill.

Distinguished from the young of *T. erosa* by its much more convex whorls and prominent ribs. The operculum appears not to be fimbriated at its edges.
  From fishes.

Cape Cod to Grand Manan Island.

5. T. AREOLATA, Stimpson.
  Shells of New England, 35.   1851.

Shell small, subperforated, turreted, red, with four distant revolving elevated ribs; aperture effuse in front; lip acute; whorls six, convex.
  Length 5, diam. 2.5 mill.

Probably a young shell, but appears distinct from any of our species. Approaches *T. reticulata*, but the transverse ribs are more prominent, and the longitudinal ones less so than in that shell. Fifteen to fifty fathoms.

Massachusetts Bay.

This is a doubtful species.

Genus **VERMETUS**, Adanson.
  Hist. Nat. Senegal, 160.   1757.

1. V. RADICULA, Stimpson.   Fig. 134.
  Shells of New England, 37.   1851.
V. lumbricalis, Gould (not Lamarck), Invert. Mass., edit. i.   1841.

Shell conic tubular, with numerous unequal raised lines or ribs along its entire length; the spiral portion consists of eight or ten closely revolving whorls, biangulate.

Length of spiral portion from one-half to one inch; the irregular prolongation or tube sometimes eight or ten inches. Animal light brown, spotted with black; mantle margin fringed; branchial plume large and long, nearly in the middle of the back; foot very short and broad, dilated into rounded auricles anteriorly; the muzzle is broad, not cleft; tongue small; tentacles short, conical; the eyes at their external bases. An elevated ridge runs along the back, becomes flattened into a membrane at the head, and passes round under the right tentacle, forming a kind of canal, near which is the anus. Operculum corneous, concentric, black, and hard on the inner, and lamellated on the outer surface.

Eggs deposited in July; soft, slightly cohering in the form of an elongated cone, bent into a half circle.

Massachusetts ; North Carolina.

Genus CÆCUM, Fleming.
Edinb. Encyc., vii. 1824.

Shell minute when young, discoidal when adult, decollated, tubular, cylindrical, arcuated; aperture round, entire; apex closed by a mammillated septum, marking the point at which the original spire has been cast off.

The animal has a long and flat rostrum; tentacles short, subclavate at the tips; eyes sessile behind the bases of the tentacles; mantle thick, fleshy, circular, closely embracing the neck; a single branchial plume; foot short, narrow, truncate in front, obtuse behind.

1. C. PULCHELLUM, Stimpson. Fig. 135.
Proc. Bost. Soc. Nat. Hist., iv. 112. 1851.

Shell clavate, arcuated, contracted at both extremities, and angular on its dorsal line; thick, pale yellowish-brown, with about twenty-five strong, rounded ribs.

Length 2.5, diam. .375 mill.

The head projects but little in advance of the foot, which is short; the muzzle is cleft and transversely wrinkled, with two black spots above, just in front of the tentaculæ, which are thick, curved, and covered with large vibrillæ; eyes conspicuous, black, oval, near the middle of the bases of the tentaculæ.

Low water to 15 fathoms.

New Bedford Harbor.

2. C. COSTATUM, Verrill.   Fig. 136.

    Amer. Journ. Science, 3d ser. iii., 210, 283, t. 6, f. 6.   1872.

C. Cooperi, S. Smith (not Carpenter), Ann. N. Y. Lyc., ix. 394, f. 3.   1870.

Shell white, moderately curved, solid, with twenty-four rounded longitudinal ribs, crossed by numerous rings, rather obscure about the middle of the shell, but very distinct at the two extremities, where the longitudinal ribs become indistinct.   There is a slight constriction near the mouth of the shell, which swells out again beyond it.   Plug mucronate, with the apex inclining to the left; operculum concave.

    Length 3, diam. 7 mill.

New England ; New York.

Genus SCALARIA, Lamarck.

Syst. Anim., 88.   1801.

The animal of the "Wentle-trap" has a retractile, proboscis-like mouth; tentacles close together, long and pointed, with the eyes near their outer bases; mantle-margin simple, with a rudimentary siphonal fold; foot obtusely triangular, with a fold in front. Exudes a purple fluid when molested.   Range from low water to 80 fathoms.

Nearly one hundred species have been described, chiefly from tropical seas.

1. S. Nov-Angliæ, Couthouy.   Fig. 137.

    Bost. Journ. Nat. Hist., ii. 96, t. 3, f. 5.   1838.

Shell with the whorls scarcely in contact; whorls ten, crossed by about eleven delicate ribs, each forming a little spine in the suture above; intervening spaces with numerous minute revolving lines; umbilicus small; glossy white.

    Length 17.5, diam. 6.25 mill.

New England.

2. S. LINEATA, Say.   Fig. 138.

    Journ. Philad. Acad., ii. 242.   1822.

Shell brownish or white, elongated, with about seven volutions ; ribs robust, obtuse, little elevated, and from seventeen to nineteen on the body-whorl; the body-whorl with generally a blackish, more or less dilated line, which is nearly concealed on the volutions of the spire by the suture; margin of the mouth robust, white, dilated below.

New England to North Carolina.

3. S. GRŒNLANDICA, Chemnitz.   Fig. 139.

S. subulata, Couthouy, Bost. Journ. Nat. Hist., ii. 93, t. 3, f. 4.   1838.
S. planicosta, Kiener, Iconog., t. 7, f. 21.   1838.

Shell tapering to a fine point, imperforate; whorls nine or ten, contiguous, slightly convex, with eight to fifteen stout compressed oblique ribs, with intervening coarse rounded vertical ridges, and seven or eight revolving striæ; the ribs not ending abruptly at the suture, but flowing along the sutural region to the preceding ones; aperture nearly circular, bordered by a rib which is emarginate at the base; operculum horny, shining.

Length 25, diam. 8.75 mill.

Animal yellowish-gray with whitish spots; mouth rather large, rounded, corrugated.

Arctic America (Eur.).

4. S. ANGULATA, Say.   Fig. 140.

(*S. clathrus*, Linn. var.), Amer. Conch., iii. t. 27.   1831.
S. Humphreysii, Kiener, Iconog.   1838.

Shell conic, turreted, imperforate, white, immaculate; whorls six to eleven, touching each other only by the ribs, but with a very narrow interval; ribs nine to each volution, prominent, simple, a little oblique, somewhat recurved, and with a more or less obvious, obtuse angle or shoulder above, near the suture; aperture suborbicular; base a little angulated; labium distinct.

Length 15 to 20 mill.

Described by Mr. Say as a doubtful variety of *S. clathrus*, from which it is distinct.

New England to Florida.

5. S. TURBINATA, Conrad.   Fig. 141.

Journ. Philad. Acad. Nat. Sciences, vii. 263, t. 20, f. 26.   1837.

Shell with the body-whorl dilated; ribs lamellar, strong, very prominent, slightly reflected, terminating above in a prominent angle; color white.

Beaufort, North Carolina.

6. S. MULTISTRIATA, Say.   Fig. 142.

Amer. Conch., iii. t. 27.   1831.

Shell conic, turreted, tapering to an acute apex, white, immaculate, imperforate; whorls about eight, in contact; costæ regular, simple, not reflected, equidistant, moderately elevated; spaces between the costæ with very numerous, approximate, equidistant, impressed lines; suture well impressed; body-whorl with about sixteen costæ.

Length 13 mill.

Massachusetts, southwards.

7. S. BOREALIS, Beck.   Fig. 143.

Acirsa borealis, Mörch.   1841.
S. Eschrichtii, Möller.   1845.

Shell white or pale flesh-color, elongated, turreted, acute; whorls ten, convex, with numerous revolving striæ; the upper whorls with slight transverse undulations or faint costæ, which are wanting on the lower ones; last whorl slightly carinated; aperture roundish, effuse, and slightly angulated in front.

Length 19, diam. 6.5 mill.

Dredged, ten to forty fathoms.

Maine, northwards.

Family VI. LITTORINIDÆ.

Shell spiral, turbiniform or depressed, never pearly; aperture rounded; peristome entire; operculum horny, paucispiral.

Animal with a muzzle-shaped head; eyes sessile at the outer bases of the tentacles; tongue denticulated; branchial plume single; foot with a linear duplication in front and a groove along the sole; mantle with a rudimentary siphonal canal; operculum lobe appendaged.   Littoral, feeding on algæ.   Distribution universal, inhabiting sea or brackish water.

Synopsis of Genera.

Shell turbinated, thick, pointed, few-whorled; aperture rounded; outer lip acute; columella rather flattened, imperforate; operculum paucispiral.

LITTORINA, Ferussac.

Shell obicular, depressed; umbilicus wide and deep; aperture rhombic; peristome thin; operculum horny, subspiral.

ARCHITECTONICA, Bolten.

Shell turbinated, thin; aperture semilunar; columella flattened, with an umbilical fissure; operculum paucispiral.   LACUNA, Turton.

Shell minute, thick, white or horny; conical, pointed, many-whorled; smooth, ribbed or cancellated; aperture rounded; peristome entire, continuous; outer lip slightly expanded and thickened; operculum subspiral.

RISSOA, Frémenville.

Shell minute, *thin*, subglobose or conical, transparent; peristome *thin*, entire; operculum annular, regular, with an internal process.

RISSOELLA, Gray.

Shell minute, orbicular, depressed, few-whorled; peristome continuous, entire, round; operculum paucispiral. **SKENEA**, Fleming.

Shell minute, discoidal, convex above, concave beneath, umbilicated; surface glossy; operculum thin, flexible, pellucid.

COCHLIOLEPIS, Stimpson.

Genus LITTORINA, Ferussac.

The periwinkles are found on the sea-shore in all parts of the world; they can exist for a lengthened period out of water, and usually inhabit situations which are only covered by the sea at high tide. The species are numerous, and the genus is represented in all parts of the world.

1. L. DILATATA, d'Orbigny. Fig. 144.

   Moll. Cuba, 207, t. 14, f. 20–23.

Shell subpyramidally conical, rather thick, imperforated, livid-gray, encircled with white nodules; whorls slanting at the upper part; columella broadly concavely dilated, purplish-brown.

Chiefly remarkable for the broadly excavated purple-brown columella, and conspicuous white nodules upon a livid-gray ground.

Beaufort, North Carolina to West Indies.

2. L. RUDIS, Donovan. Fig. 145.

   (*Turbo.*) · Brit. Shells, i. t. 33, f. 3. 1800.

T. obligatus, Say, Journ. Philad. Acad., ii. 241. 1822.

Shell very strong and coarse, subovate, ventricose; whorls five to six, convex, tapering rapidly to a little elevated spire, and covered with revolving elevated lines and grooves; body-whorl with ten to twelve revolving costæ, the intervening spaces finely reticulated; lip plaited by the termination of the costæ; about four of these on the next whorl, and obsolete above; base of the lip broadly bevelled; pillar-margin also broadly flattened; aperture regularly oval; color obscurely brownish, sometimes orange or olive, occasionally banded with white.

Length 12.5 mill.

New England and Middle States (*N. Europe*).

3. L. TENEBROSA, Montagu. Fig. 146.

   (*Turbo.*) Test. Brit., 303, t. 20, f. 4. 1803.

T. vestitus, Say, Journ. Philad. Acad., ii. 241. 1822.

L. rudis (part), Stimpson, Shells N. E., 33. 1851.

Shell small, conic, not as stout as *rudis;* spire elevated and pointed, as long as the aperture; whorls five to six, rounded, with faint revolving lines; suture deeply impressed; lip thin, acute; color black, brown, green, or reddish, sometimes reticulated or striped with colored lines.

Length 12.5, diam. 7.5 mill.

Animal with a dark olive head, and an olive stripe on the tentacles from the eye; sides of the foot lined with the same.

New England and Middle States (*N. Europe*).

4. L. LITOREA, Linnæus. Fig. 147.

(*Turbo.*) Syst. Nat., edit. xii. 1232.

T. ustulatus, Lam., Anim. s. Vert., edit. Deshayes, ix. 214.

L. vulgaris, Sowb., Genera of Shells. Littorina, f. 1.

Shell ovately turbinated, imperforated, thick, smooth or with elevated spiral striæ; whorls sometimes concavely impressed round the upper part; olive, ash, or red, sometimes banded and lineated with black; columella broadly callous, slightly excavated, white.

New England (*N. Europe*).

5. L. PALLIATA, Say. Fig. 148.

(*Turbo.*) Journ Philad. Acad. Nat. Sci., ii. 240. 1822.

?T. neritoides, Linn., Syst. Nat., edit. xii. 1232. 1767.

L. littoralis, Forbes and Hanley, Brit. Moll. 1853.

Shell semiglobose, very solid, spire flatly depressed; whorls obliquely convex, smooth or very obscurely striated; yellow, sometimes broadly brown-banded; aperture circular, very much contracted; columella broadly excavated.

Well distinguished by its oblique, obtuse growth and depressed spire, varying in color from yellow, more or less banded, to freckled-brown.

New England and Middle States (*N. Europe*).

6. L. IRRORATA, Say. Fig. 149.

(*Turbo.*). Journ. Philad. Acad., ii. 239. 1822.

Shell solid, robust, pyramidal, with numerous, elevated. obtuse, equal lines; suture not indented; spire acute; pillar-lip thickened; lip stout, bevelled to a moderately thin edge, which is everted below; aperture oval, angulated above; color pale ash or cinereous or deep brown; pillar-lip umber-brown; lip on its margin with purple abbreviated lines.

Length 25, diam. 13 mill.

Whole Coast.

Reeve's figure (Icon. x. f. 56) does not represent this species, and it does not occur at Sitka as stated by him.

Genus **ARCHITECTONICA**, Bolten.

Mus. Bolt. 1798.

Solarium, Lamarck, Prodr., 74. 1801.

The species of this genus are fancifully called "stair-case shells" from the appearance of the spiral edges of the whorls in the perspective umbilicus. Distribution tropical.

1. A. GRANULATA, Lamarck. Fig. 150.

(*Solarium.*) Anim. s. Vert., vii. 3. 1822.

Shell conoid, yellowish flesh-color, stained with livid-purple, sparingly belted with distant chestnut-red spots and dots; whorls spirally grooved and granosely warted; base many crenated; umbilicus rather small.

North Carolina to West Indies.

Genus **LACUNA**, Turton.

Zool. Journ., iii. 190. 1827.

The Lacunæ feed upon sea-weed, and Lovén observes that when the fuci are of a brown color these animals become green, but if red they assume a rosy tint. They principally inhabit the shores of northern countries, and several species are common to both continents.

1. L. DIVARICATA, Fabricius. Fig. 151.

(*Turbo.*) Fauna Grœnl., 392. 1780.

T. vincta Montagu, Test. Brit., 307, t. 20, f. 3. 1803.

T. quadrifasciatus, Fleming, Brit. Anim., 299. 1828.

L. pertusa, Conrad, Journ. Philad. Acad., vi. 266, t. 11, f. 19. 1830.

Shell small, thin, ovate-conic; spire pointed; whorls five, very convex, with faint incremental lines; suture deep; aperture nearly circular; lip sharp and simple; pillar-lip with a wide and deep groove behind, ending in a profound umbilicus; color yellowish, with sometimes four or five dark purplish or reddish bands. Length 7.5 mill.

New England to New York. (*England.*)

2. L. NERITOIDEA, Gould. Fig. 152.

Amer. Journ. Science, xxxviii. 197. 1840.

? L. *pallidula*, Turton (var.), Zool. Journ., iii. 190. 1827.

Shell small, thin, hemispherical, or obliquely ovate; whorls three and a half, regularly convex, minutely wrinkled near the suture, and with an occasional transverse scratch, otherwise

6

smooth, and covered with a rough, greenish-yellow epidermis; sutural region depressed and sub-channeled; the spire scarcely prominent above the very large lower whorl, and placed a little to one side; aperture oblique, semicircular.

Length 5, diam. 6 mill.

New England.

Genus RISSOA, Frémenville.

Desmarest, Bull. d. Sc., par la Soc. philom., 7.   1814.

Cingula, Fleming, Brit. Anim., 207, 305.   1828.

The animal has large, slender tentacles, with eyes on small prominences near their outer bases; the foot is pointed behind; the operculigerous lobe has a wing-like process and a filament on each side.

Universally distributed, but principally in the north temperate zone. They range from high water to one hundred fathoms, but abound most in shallow water, near the shore, on beds of fucus and zostera.

1. R. MINUTA, Totten.   Fig. 153.

(*Turbo.*)   Amer. Journ. Science, xxvi. 369, f. 7.   1834.

Shell minute, conic, thin, polished, elevated to an obtuse apex; whorls five, convex, with very fine transverse striæ; suture distinct, with a round shoulder on the whorl; aperture oval, entire, rounded at the base, very slightly angular above; lip sharp; lower portion of the pillar-lip slightly recurved, with a loosely attached enamel which rises before an umbilical pit; operculum horny, sub-spiral; yellowish-brown, usually covered by a dark green pigment.

Length 3.75, diam. 1.5 mill.

Animal dusky-brown; tentacles, and a line on each side the neck, light drab. Very active in movement.

New England.

2. R. LATIOR, Mighels and Adams.   Fig. 154.

(*Cingula.*)   Bost. Journ. Nat. Hist., iv. 48, t. 4, f. 22.   1844.

Shell minute, ovate-conic, smooth, pale horn-color; whorls more than four, convex; suture much impressed; last whorl broad, larger than the rest of the shell; aperture ovate-orbicular, left margin with a lamina.

Length 2, diam. 1.25 mill.

Maine.

3. R. ROBUSTA, H. C. Lea. Fig. 155.
(*Cingula.*) Bost. Proc. 288, t. 24, f. 4. 1844.

Shell ovate-acuminate, perforated, smooth, thick, white; spire short, subacute; suture impressed; whorls five, somewhat angled at the superior suture; last whorl round; base smooth; perforation narrow, profound; mouth ovate, large.

Length 2.5, diam. 2 mill.

Cape May, N. J.

This is a somewhat doubtful species: it (as well as the next) is perhaps only the young of a form of *R. latior*. Only one specimen was obtained.

4. R. MODESTA, H. C. Lea. Fig. 156.
(*Cingula.*) Bost. Proc. 288, t. 24, f. 5. 1844.

Shell ovate, imperforate, smooth, thin, diaphanous, greenish horn-color; spire short, ovate, not acute; suture small; whorls four, flattish; last whorl round; base smooth; mouth ovate; acute above, rounded below.

Length 2.5, diam. 1.8 mill.

Quite common on the under surface of stones below high-water mark.

Long Island (near Brooklyn).

5. R. TURRICULA, H. C. Lea. Fig. 157.
(*Cingula.*) Proc. Bost. Soc. Nat. Hist. 289, t. 24, f. 6. 1844.

Shell elevated, conic, perforate, smooth, thick, tawny; spire very much exserted, conical, obtuse, sutures small; whorls six, convex; last whorl slightly bullate; base smooth; perforation small, narrow, lunate; mouth ovate; columella thick, almost disjoined from the last whorl.

Length 3, diam. 1.3 mill.

South Carolina.

This species has not been detected or identified by subsequent collectors.

6. R. ACULEUS, Gould. Fig. 158.
(*Cingula.*) Invert. Mass. Edit. 1, 266, f. 172. 1841.

Shell minute, subcylindrical, elongated, fragile; whorls six, very convex, with a deep suture; with numerous revolving lines and traces of longitudinal folds towards the apex. Aperture small, suboval, oblique. Light horn-color.

Length 5 mill.

The animal is white, with moderately produced head and foot slightly dilated at the anterior angles; eyes black.

New England. (*Europe.*)

According to Mr. Gwyn Jeffreys, this species = *striata*, J. Adams, 1795.

7. R. MULTILINEATA, Stimpson.   Fig. 159.
   Proc. Bost. Soc. Nat. Hist. iv. 14.   1851.

Shell minute, oblong-ovate, blunt, white; whorls five, convex, marked with about twenty minute, transverse striæ; aperture orbicularly ovate, peristome not thickened, effuse.

Shorter than *aculeus*, with the whorls more compactly coiled and stronger striæ. The latter are more numerous than in the following species.

Length 2.5, diam. 1 mill.

New England.

8. R. MICHELSI, Stimpson.   Fig. 160.
   Proc. Bost. Soc. Nat. Hist. iv. 15.   1851.
   Cingula arenaria, Mighels and Adams (not Montagu).   Bost. Journ. iv.
       49, t. 4, f. 24.   1842.

Shell minute, white, subcylindrical, subplicate longitudinally and minutely striate transversely; spire elongated, conical; whorls six, convex; suture impressed; aperture moderate, suborbicular.

Length 2.5, diam. 1.2 mill.

Maine.

9. R. EXARATA, Stimpson.   Fig. 161.
   Proc. Bost. Soc. Nat. Hist. iv. 15.   1851.

Shell small, ovate, fuscous, rather solid, imperforate; whorls five, rather convex, subplicate posteriorly, and with inequidistant, elevated, transverse ribs, three on the upper whorls. Aperture small, ovate, peristome thickened.

Length 2.6, diam. 1.2 mill.

Massachusetts.

10. R. CARINATA, Mighels and Adams.   Fig. 162.
    (*Cingula.*)   Bost. Journ. Nat. Hist. iv. 49.   1842.
    Cingula semicostata.   Mighels and Adams (not Montagu).   Bost. Journ.
        iv. 49, t. 4, f. 23.   1842.
    Rissoa pelagica, Stimpson, Bost. Proc. iv. 15.

Shell very small, ovate conical, ferruginous, very thin; whorls five, convex, the upper ones with longitudinal ribs, the lower half of the body whorl with revolving striæ. Aperture nearly orbicular, peristome thin and sharp.

Length 2.6, diam. 1.7 mill.

Maine, northwards.

11. R. LÆVIS, De Kay.   Fig. 163.

(*Cingula.*)   Moll. New York, 111, t. 6, f. 118.   1843.

Shell small, elevated, moderately solid; whorls five, very convex, with deep sutures; surface smooth; body-whorl large; aperture small, oval, the columellar lip partially everted over the rather large umbilicus.   White.

Length 5 mill.

Connecticut.

This is a doubtful species.

12. R. PATENS, Gould.

Bost. Proc. viii. 280.   1862.

Shell minute, ovate, thin, smooth; whorls five or six, ventricose, with a subsutural impressed line; aperture rounded, emarginate posteriorly; columella but slightly reflexed; lips somewhat thickened, fuscous.

Length 3, diam. 2 mill.

Remarkable for its large aperture and subsutural impressed line.

Fort Johnson, Charleston Harbor, S. C.

13. R. INCOMPTA, Gould.

Bost. Proc. vii. 280.   1862.

Shell small, elongated, vitreous, reddish-white; whorls seven, rounded, with three revolving ribs and longitudinal lines, the body-whorl carinate; aperture circular, the peristome simple, thickened.

Length 2, diam. 1 mill.

(*Coral Sand*) *Florida.*

Genus RISSOELLA, Gray.

Zool. Proc. 159.   1847.

This genus differs from *Rissoa* in the shells being thin and without thickened lip; the operculum differs also, in being annular with a central internal process.

The animals are found adhering to floating sea-weeds, in pools between tide-marks; their eyes are situated so far behind on the head, that the transparency of the shells appears to be essential to the vision of the animal.

1. R. EBURNEA, Stimpson.   Fig. 164.

(*Rissoa.*)   Proc. Bost. Soc. Nat. Hist. iv. 14.   1851.

Shell small, ovate-conoid, white, shining, smooth; whorls four, rather convex, subangulated at the suture; aperture ovate-elliptic; peristome thin, simple, acute, effuse anteriorly.

Length 4.5, diam. 2.2 mill.

Massachusetts.

2. R. SULCOSA, Mighels.   Fig. 165.

(*Phasianella.*)   Bost. Journ. Nat. Hist. iv. 348, t. 16, f. 4.   1843.

Shell very small, ovate-conical, smooth and white; whorls four, slightly convex, with six or seven transverse grooves on the body-whorl, and three on each of the two next above, spire smooth and pointed; aperture ovate-oblong, with three slightly apparent transverse bands within, as seen under a strong magnifying power.

Length 2.5, diam. 1.3 mill.

Maine.

Genus SKENEA, Fleming.

Hist. Brit. Anim. 297, 313.   1828.

1. S. PLANORBIS, Fabricius.   Fig. 166.

(*Turbo.*)   Fauna Grœnl. 394.   1780.

Shell minute, flat, slightly convex above, broadly concave below, with a perspective umbilicus; whorls three, smooth, light horn-color; aperture small, circular, with sharp lip.   Operculum multispiral, horny.

Height .8, diam. 1.2 mill.

Clinging to stones at low-water mark.

Mass., northwards.   (*N. Europe.*)

Genus COCHLIOLEPIS, Stimpson.

Bost. Proc. vi., 308.   1858.

1. C. PARASITICUS, Stimpson.   Fig. 167.

Bost. Proc. vi., 308.   1858.

Shell thin, discoidal, convex above, concave and umbilicated below; the edge thin and sharp; whorls three, rapidly enlarging; surface smooth and glossy; lip not thickened.   Operculum thin, flexible and pellucid.

Animal blood red; foot oblong, tapering behind, and slightly emarginate in front; head small, rounded, with long, slender, tapering tentacles.   Eyes none?

Parasitic on *Acoëtes lupina.*

Charleston, S. C.

Family TURBINIDÆ.

Shell spiral, turbinated or pyramidal, *nacreous inside*. Operculum calcareous and paucispiral, or horny and multispiral.

Animal with a short muzzle; eyes pedunculated at the outer bases of the long and slender tentacles; head and sides ornamented with fringed lobes and tentacular filaments.

Distribution universal; feeding on sea-weeds.

The shells are brilliantly pearly under the epidermis and within the aperture.

Synopsis of Genera.

Shell turbinated, solid; whorls convex, often grooved or tuberculated; aperture large, rounded, slightly produced in front; operculum shelly and solid, callous outside, and smooth, or variously grooved and mammillated, internally horny and paucispiral. TURBO, Linnæus.

Shell thin, globosely depressed, whorls convex, smooth or transversely striated; aperture nearly circular; columella ending in a simple point. MARGARITA, Leach.

Shell minute, *not nacreous*, depressed, few whorled, deeply umbilicated; peristome entire, nearly continuous, sinuated on its inner side, and slightly so externally; operculum shelly, multispiral. ADEORBIS, S. Wood.

Genus TURBO, Linnæus.

Syst. Nat. Edit., x. 761. 1758.

1. T. CRENULATUS, Gmelen. Fig. 168.

Syst. Nat., 3575. 1790.

Shell ovate, imperforated, sutures of the spire excavated, whorls covered with papillose nodules, convex or slightly angulated in the middle, and ridged, ridges generally squamose, the upper scales being more prominent and erect; operculum thick, testaceous.

Whitish, with rays or blotches of fawn-color or reddish, pearly within.

North Carolina to West Indies.

Genus MARGARITA, Leach.

Journ. de Phys., lxxxviii., 464. 1819.

The species of this genus are principally inhabitants of northern or antarctic seas. They are generally more depressed, smoother, and smaller than *Turbo*, which is tropical in distribution.

1. M. OCCIDENTALIS, Mighels and Adams. Fig. 169.
   (*Trochus*.) Bost. Journ. Nat. Hist., iv. 47, t. 4, f. 16. 1842.
Margarita alabastrum, Beck.

Shell small, rather solid, subtranslucent, pale horn-color, with light brown revolving carinæ, three in number on the upper whorls, and from four to six on the body-whorl; whorls seven, convex, with distinct sutures; spire small; the body-whorl large, with coarse revolving striæ around the indented umbilical region.
Height 12.5, diam. 10 mill.

Maine. (*Eur.*)

2. M. CINEREA, Couthouy. Fig. 170.
   (*Turbo.*) Bost. Journ. Nat. Hist., ii. 99, t. 3, f. 9. 1838.

Shell small, thin, pyramidal; whorls five to seven, with several revolving ribs, the central one largest; umbilicus broad and deep; lip sharp; aperture circular, slightly reflected over the umbilicus. Ashen-gray or greenish.
Height 12.5, diam. 10 mill.

New England. (*Eur.*)

3. M. OBSCURA, Couthouy. Fig. 171.
   (*Turbo.*) Bost. Journ. Nat. Hist., ii. 100, t. 3, f. 2. 1838.

Depressed, conical, solid; spire obscure, reddish-brown, base ash-colored; whorls angulated by two or three revolving ridges; lines of growth coarse; aperture circular; pearly within.
Height 10, diam. 15 mill.

New England. (*Eur.*)

4. M. VARICOSA, Mighels and Adams. Figs. 172, 178.
   Bost. Journ. Nat. Hist., iv. 46, t. 4, f. 14. 1842.
Margarita acuminata, Mighels and Adams, ib. f. 15. 1842.

Shell small, conical, thin, dingy white or drab-color; whorls four, convex, with numerous longitudinal oblique ribs, and crowded revolving striæ. Suture distinct, subcanaliculate; umbilicus rather large and deep, bounded by two rather rugged varices. Aperture circular, lip sharp.
Height 6.25, diam. 6.25 mill.

New England, northwards. (*Eur.*)

M. acuminata is the young of this species. Fig. 178 represents it.

5. M. MINUTISSIMA, Mighels. Fig. 173.
   Bost. Journ. Nat. Hist., iv. 349, t. 16, f. 5. 1843.

Shell very minute, subovately globose; whorls three, convex, longitudinally furrowed ; spire short, obtuse ; suture strongly impressed, aperture orbicular ; umbilicus large, deep. Dull ash-color.

Height 5, diam. 5 mill.

Casco Bay, Maine.

6. M. UNDULATA, Sowerby.   Fig. 174.

Malacol and Conch. Mag. i., 26.

Margarita striata, var. *Grœnlandica,* Möller, Ind. Moll. Grœn. 1842.
Turbo incarnatus, Couthouy, Bost. Journ. Nat. Hist., ii. 98, t. 3, f. 13. 1838.
Trochus tumidus, Montagu, Leth. Suev. t. 30, f. 3.

Shell orbicular, small, smooth and shining ; whorls four or five, convex, with numerous striæ, alternately finer and undulated near the sutures by short folds or wrinkles ; umbilicus quite large and deep ; aperture nearly circular, very oblique.   Uniform red of various shades.

Height 7.5, diam. 10 mill.

New England. (Eur.)

Trochus Grœnlandicus, Chemnitz, 1781, may be this species.

7. M. HELICINA, Fabricius.   Fig. 175.

(*Turbo.*)   Fauna Grœnl. 1780.

Margarita arctica, Leach, Ross' Voyage. 1819.
Turbo inflatus, Totten, Am. Journ. Science, xxvi. 368, f. 5. 1834.

Shell small, thin, translucent, shining and globular ; whorls five, convex, with revolving minute lines on the base ; spire low, convex ; suture impressed ; aperture large, circular and expanded ; umbilicus large and deep.   Pale horn-color.

Height 5, diam. 6.25 mill.

New England.

8. M. CAMPANULATA, Morse.   Fig. 176.

Shell small, depressed, orbicular, smooth, shining, translucent ; spire minute, pointed, aperture large, companulate.   Light olive or horn-color.

Length 3.5, diam. 7 mill.

New England.

9. M. ARGENTATA, Gould.   Fig. 177.

Invert. Mass. Edit. i. 256, f. 174. 1841.

Trochus glaucus, Möller, Ind. Moll. Grœnl. 1842.

Shell minute, conical, with an obtuse tip ; pearly white ; whorls four, convex, the last slightly angular, covered with fine, crowded, revolving lines ; suture deep ; aperture circular.

Height 2.5, diam. 3 mill.

New England. (Europe.)

Doubtful species.

10. M. ORNATA, De Kay.   Fig. 179.

    Moll., New York, 107, t. 6, f. 104.   1843.

Shell moderately solid, subconical; its transverse exceeding its vertical diameter; whorls four to five, convex; the body-whorl very large, subinflated; seven to nine distant revolving costæ on its upper surface, which is separated from the simply striate surface beneath by an obsolete carina; spire not much elevated, faintly striate; umbilicus large and very profound; aperture rounded, oblique; lip thin and simple, entire.   Bright red.

    Length 2.5, diam. 3.75 mill.

New York.

11. M. MULTILINEATA, De Kay.   Fig. 180.

    Moll., New York, 109, t. 6, f. 108.   1843.

Shell small, pyramidal; whorls four, convex, obtusely carinate; suture impressed; spire elevated; whorls with minute revolving striæ, and three to four revolving ribs; aperture suborbicular; umbilicus entirely concealed by the reflection of the lip, but its place marked by a slight depression. Beautifully variegated by alternate yellowish-white and brown, or reddish-brown revolving lines; lip with abbreviated red and white lines.

    Height 7.5 mill.

New York.

Genus **ADEORBIS**, S. Wood.

Ann. Mag. Nat. Hist. 530.   1842.

1. A. COSTULATA, Möller.   Fig. 181.

    Index Moll. Grœn. 8.   1842.

Shell minute, white, thin, with crowded longitudinal ribs and fine revolving striæ on the base; umbilicus deep; aperture rounded with continuous peristome; operculum multispiral, of about eight volutions, the outer ones testaceous, the nuclear corneous.

    Diam. 2.5 mill.

New England, northwards. (Eur.)

Family JANTHINIDÆ.

The animal has a proboscidiform head; tentacles short and obtuse, with pointed eye-pedicels at their bases, but without any trace of eyes. Gills plumose, partially exserted. Foot small, flat, rudimentary, furnished with a vesicular appendage on the hinder part. Pelagic.

The shell is very thin, translucent, with a sinistral nucleus. Violet colored. The family contains but one genus.

Genus JANTHINA, Bolten.
Mus. Bolt. 1798.

1. J. FRAGILIS, Bruguiere. Fig. 182.

Helix Janthina, Gmelin, Syst. Nat. 3645. 1790.
Janthina communis, Lam. Anim. s. Vert., vii. 206. 1822.

Shell depressly semiglobose, flattish beneath; whorls slopingly convex, rather rudely decussately striated; whitish above, violet below, whitish around the columella; aperture transverse; a little sinuated in the middle.

Driven upon the Northern Shores during storms.

Family FISSURELLIDÆ.

Animal with a well-developed head, a short muzzle, subulate tentacles, and eyes on rudimentary pedicels at their outer bases; sides ornamented with short cirri; branchial plumes two, symmetrical; anal siphon occupying the anterior notch or perforated summit of the shell.

Synopsis of Genera.

Shell oval, conical, depressed, with the apex in front of the centre, and perforated; surface radiated or cancellated; muscular impression with the points incurved.   FISSURELLA, Bruguiere.

Shell conical, elevated, with the apex recurved; perforation in front of the apex, with a raised border internally; surface cancellated.
                    CEMORIA, Leach.

Genus FISSURELLA, Bruguiere.
Encyc. Meth. i., 14. 1789.

1. F. ALTERNATA, Say. Fig. 183.

Journ. Philada. Acad. Nat Sci., ii. 224. 1822.

Shell oblong-ovate, moderately thick, cinereous or dusky, with equal concentric lines crossed by alternately larger and smaller radii, all of which are equable or not dilated in any part; vertex placed nearer the smaller end; perforation oblique, oblong, and a little contracted in the middle; within white; margin simply crenate; apex with an indented transverse line at the larger end of the perforation.

Height 10, length 20, diam. 15 mill.

Southern Coast.

Genus CEMORIA, Leach.

Lowe, Zool. Journ., iii. 76. 1826.

Rimula, Lovén (not Defrance), Ind. Moll. Scand. 21. 1846.
Puncturella, Lowe, Zool. Journ. iii. 78. 1827.

In the second edition of Gould's Invertebrata of Massachusetts, occurs an error in the description of this genus which it is necessary to correct: the apex is said to be "curved forwards, with a fissure just behind the apex." This is reversing the real position, as the apex is recurved, with the fissure in front. See the illustration of shell, with animal, in H. and A. Adams' "Genera," t. 51, f. 7.

1. C. NOACHINA, Linnæus.   Figs. 184, 185.
    (*Patella.*)   Mantissa, 551. 1771.

Patella aperta, Montagu, Test. Brit. 491, t. 13, f. 10. 1803.
Patella fissurella, Müller, Zool. Dan. i., t. 24, f. 4-6. 1788.
Cemoria Flemingii, Leach, Sowerby, Conch. Man., f. 244. 1842.
Sipho striata, Brown, Brit. Conch. t. 36, f. 14-16. 1844.
Cemoria princeps, Mighels and Adams, Bost. Journ. iv. 42, t. 4, f. 3. 1842.
Diodora Noachina, Stimpson, Shells, N. E. 30. 1851.

Shell small, conical; apex recurved, obliquely perforated; surface covered by about twenty unequal radiating ribs which feebly crenate the margin.

Height 2.5, length 5 mill.

N. of Cape Cod, Mass. (N. Eur.)

Family CALYPTRÆIDÆ.

Shell limpet-like, with a more or less spiral apex; interior (in our species) divided by a shelly partition to which the adductor muscles are attached.

The animal has a lengthened muzzle; eyes on the external bases of the tentacles; branchial plume single. They feed on sea-weed, and live attached to the surface of rocks or other shells, their forms modified to conform to the situation they inhabit.

Synopsis of Genera.

Shell subconic, spiral; apex subcentral; aperture wide, with the internal appendage entire and cup-shaped, attached by one of its sides.

CRUCIBULUM, Schum.

Shell ovate or oblong ; apex posterior, oblique, submarginal; aperture elongated, polished within, the posterior half covered by a horizontal testaceous lamina.　　　　　　　　　　　　　CREPIDULA, Lam.

Genus CRUCIBULUM, Schumacher.

Essai d'un Nov. Syst. 182. 1817.

1. C. STRIATUM, Say. Fig. 186.

(*Dispotæa.*) Journ. Philada. Acad. Nat. Sci. v. 216. 1826.

Shell moderately solid, conical, with numerous equidistant, elevated radiating lines. Summit smooth, obtusely pointed, subspiral, inclining towards the left side and posterior end. Internal cup attached at one side, and terminating above near the inner apex of the shell. White.

Height 12.5, diam. 20 mill.

New England to New Jersey.

Undetermined Species.

INFUNDIBULUM DEPRESSUM, Say.

Journ. Acad. Nat. Sci., Phila., v. 209. 1826.

Shell depressed, fragile, with small concentric irregular wrinkles; volutions three; suture not profoundly indented; apex not central; base oval, almost orbicular; umbilicus oblong; internal plate small.

Diameter 5 + mill.

South Carolina.

Genus CREPIDULA, Lamarck.

Prodromus. 1799.

1. C. FORNICATA, Linnæus. Figs. 187, 188, 189.

(*Patella.*) Syst. Nat. Edit., xii. 1257. 1767.

Crepidula glauca, Say, Journ. Philad. Acad., ii. 226. 1822.
Crepidula convexa, Say, ibid., 227. 1822.

Shell varying in convexity, with one side more oblique than the other; apex turned to one side ; surface transversely wrinkled. Partition smooth, slightly concave. White, or greenish, or reddish, with longitudinal undulated chestnut-colored lines, sometimes broken up into spots.

Length 1 to 2 inches, width .7 to 1.3 inch.

Inhabits the entire coast. (*Eur.*)

C. *glauca* (Fig. 189) is the young shell when flattened ; if the growth is normally convex. however, the young is the *C. convexa* (Fig. 188). I have satisfied myself that these two forms are both

juveniles of *C. fornicata*, by the examination of numerous specimens.

2. C. UNGUIFORMIS, Lamarck.   Fig. 190.

Anim. s. Vert., vi. 25.   1819.
Crepidula plana, Say, Journ. Philad. Acad. Nat. Sci., ii. 226.   1822.

Shell subovate or quadrilateral, depressed, concave, from generally inhabiting the interior of the mouth of univalve shells, surface wrinkled; white.

Length 1 to 1.5 inch, breadth .7 to 1 inch.

This species is generally parasitic on other shells, and prefers
the interior of Naticas, Busycons, etc., attaching itself just within
the aperture.   It has been supposed by Gray and others that it
is not a distinct species, but merely the *C. fornicata*, modified in
color and form by situation.   This idea is incorrect, because I
have collected *C. unguiformis* from *external* surfaces, yet it still
retains its plain white color, and is always as nearly flat as cir-
cumstances will permit.

Inhabits the entire coast.

3. C. ACULEATA, Gmelin.   Fig. 191.

(*Patella.*)   Syst. Nat., 3693.   1790.

Shell ovate, laterally incurved at the apex, radiately irregularly
ribbed, ribs with tubercles or vaulted scales, sometimes growing
into short spires; brownish, sometimes rayed, brown within.

Southern Coast.

Family PATELLIDÆ.

The limpets have a conical shell with a non-spiral apex, not
perforated; muscular impression horse-shoe shaped.   The animal
has a distinct head, furnished with tentacles, bearing eyes at their
outer bases; foot as large as the margin of the shell; mantle
plain or fringed.   Respiratory organ in the form of one or two
branchial plumes, lodged in a cervical cavity, or of a series of
lamellæ surrounding the animal between its foot and mantle.
Mouth armed with a horny upper jaw, and a long ribbon-like
tongue furnished with numerous teeth.

The species are very numerous, and distribution universal.

Genus PATELLA, Linnæus.

Syst. Nat., edit. x. 780.   1758.

The shells described under this genus have been assigned, the
first to the genus *Lepeta*, the others to *Tectura* by modern authors,

upon differences in the animals which do not appear to me to be of generic value.

1. P. CÆCA, Müller. Fig. 192.

Zool. Danica, i. 45, t. 12, f. 1, 2, 3. 1788.

Patella candida, Couthouy, Am. Journ. Science, xxxiv. 217. 1838.

Shell small, conical, with numerous minute revolving ribs crossed by fine concentric lines, giving the surface under the lens the appearance of network; summit nearly central; margin slightly scolloped by the termination of the ribs. White.

Length 8.75, height 2.5 mill.

New England, northwards. (*Eur.*)

2. P. TESTUDINALIS, Müller. Fig. 193.

Prod. 237.

Patella tessellata, Müll., Zool. Dan. Prodr., iii. 2868. 1788.

Patella Clelandi, Sowerby, Trans. of Linn. Soc., viii. 621.

Patella virginea, Müll., Zool. Dan. Prodr., iii. 2867. 1788.

Patella amœna, Say, Journ. Philad. Acad. ii. 223. 1822.

Patella clypeus, Brown, Brit. Conch., t. 37, f. 9, 10. 1827.

Lottia antillarum, Sowerby, Conch., Man. f. 231.

Shell oblong-oval, frequently with a calcareous deposit, under which we observe numerous radiating lines, which are crossed by minute concentric wrinkles. Margin entire, acute; apex behind the middle, and turning towards the short end. Whitish or greenish, with brown bands, frequently interrupted, forming square tessellated spots; within bluish-white, etc., with an apicial brown spot and marginal band.

Length 20 to 38, width 12 to 20 mill.

Northern Coast. (*Europe.*)

3. P. ALVEUS, Conrad. Fig. 194.

Journ. Philad. Acad., vi. 267, t. 11, f. 20. 1831.

Shell oblong, sublinear, elevated, thin, pellucid, with fine radiating striæ and fine concentric lines; sides nearly straight; apex not central, pointing to the short end. Whitish, with reddish-brown spots and lines, visible within.

Length 7.5 to 12.5, width 5 to 7.5 mill.

New England.

This is doubtfully distinct from *testudinalis*.

Family DENTALIDÆ.

Genus DENTALIUM, Linnæus.

Syst. Nat., edit. x. 785. 1758.

The "tooth shells" are tubular, symmetrical, curved, open at each end, attenuated posteriorly; surface smooth or longitudinally striated; aperture circular, not constricted.

The animal is attached to its shell near the posterior anal orifice; head rudimentary, no tentacles or eyes; oral orifice fringed; foot pointed, conical, with symmetrical side lobes, and an attenuated base, in which is a hollow communicating with the stomach. Branchiæ two, symmetrical, posterior to the heart; blood red! Sexes united? Tongue denticulate.

These anomalous animals are animal feeders; they live in all seas, ranging from ten to one hundred fathoms.

1. D. DENTALE, Linnæus.   Fig. 195.

Syst. Nat., edit. xii. 1263. 1767.
Dentalium striatum, Montagu, Test. Brit., 435. 1803.
Dentalium attenuatum, Say, Journ. Philad. Acad., iv. 154, t. 8, f, 3. 1825.
Dentalium occidentale, Stimpson, Shells of New England, 28. 1851.

Shell slender and tapering, shaped like an elephant's tusk; the tip cut off, leaving a very small opening. Surface rather glossy, yellowish-white, marked with about twenty closely arranged unequal rib-like striæ, running the whole length of the shell.

Length 1 inch.

New England.

2. D. STRIOLATUM, Stimpson.   Fig. 196.

(*Entalis.*) Proc. Bost. Soc. Nat. Hist., iv. 114. 1851.
Dentalium entalis, Mighels (not Linn.).
Dentalium abyssorum, Sars. 1858.

Shell large, slightly curved, rugose from the growth lines, but destitute of longitudinal striations. White.

Length 2 inches.

Maine. (Eur.)

This species has been referred to the genus *Entalis*, which is said to be distinguished from *Dentalium* by the presence of a notch-like or narrow longitudinal fissure communicating with the perforated apex. Upon examination of a large number of species I find that, in some at least, this character is not even of specific value.

3. D. PLIOCENUM, Tonmey and Holmes.   Fig. 197.
Pliocene Foss. So. Car., 105, t. 25, f. 2.   1857.

Shell slightly curved, marked by about thirty-eight very indistinct ribs, which become obsolete towards the base; lines of growth numerous, indistinct; aperture orbicular.

(Living?)   *South Carolina.*

Family CHITONIDÆ.

Head surrounded by a semicircular veil or hood; eyes and tentacles none; mouth with cartilaginous jaws; gills in a series of lamellæ, between the mantle and foot round the sides and posterior part of the body; foot oblong, rounded at each end. Lingual ribbon long and linear, with numerous transverse series of teeth.

The Chitons are abnormal mollusks in many respects. Their shell in eight separate but connected pieces gives them an articulated appearance; the heart is central; the reproductive organs symmetrical, with two orifices, and the sexes united; the intestine is straight, and the anal orifice posterior and median.

Notwithstanding these resemblances to the annelids they are believed to be more closely related to the mollusca.

Genus **CHITON**, Linnæus.
Syst. Nat., edit. x. 1758.

1. C. MENDICARIUS, Mighels and Adams.   Fig. 198.
Bost. Journ. Nat. Hist., iv. 42, t. 4, f. 8.   1842.
C. Hanleyi, Bean. Thorpe's Brit. Mar. Conch., 263.   1844.

Shell cinereous, with dark clouds, long, oval with obtuse dorsal ridges; surface with elevated dots or granules, disposed in longitudinal lines, except towards the margin, where they are irregular and larger; no visible concentric striæ; triangular areas very indistinct, outer whorls small, margin coriaceous, red.

Length 25, breadth 10 mill.

Maine, northwards.   (*Eur.*)

2. C. APICULATUS, Say.   Fig. 199.
American Conch., No. 7.   1834.

Shell oblong-oval, convex; valves obtusely carinate, the central portion of the posterior margins becoming slightly beaked with age. Lateral areas triangular, studded with numerous rounded tubercles, obsolete towards the apices, more numerous towards

7

the lateral margins, which are rounded with an elevated marginal line. Medial areas lozenge-shaped, with numerous elevated rounded dots arranged in ten or twelve series on each side of the carina, parallel with the longitudinal axis of the body; grayish, bluish, or ferruginous.

Length 13 to 25, width 7.5 to 15 mill.

Whole Coast.

3. C. CINEREUS, Linnæus. Fig. 200.
    Syst. Nat., edit. xii. 1107. 1767.
Chiton marginatus, Pennant, Brit. Zool., iv. 61, t. 36, f. 2. 1777.

Shell small, ovate, carinate and pointed behind; surface apparently smooth, but under the lens minutely shagreened in diamond-shaped granules; dull ashen or greenish.

Length 12, width 7 mill.

New England. (Eur.)

4. C. MARMOREUS, Fabricius. Fig. 201.
    Fauna Grœnlandica, 420. 1780.
C. fulminatus, Couthouy, Bost. Journ. Nat. Hist., ii., p. 80, t. 3, f. 19. 1838.

Shell oblong-ovate, rather flat, color varying from bright red to yellowish or dark reddish-brown, with numerous fine zigzag whitish lines arranged over the whole surface, and a line of six or eight whitish spots alternating with dark red along the posterior edge of each valve; valves carinated and slightly beaked, their surface covered with microscopic granulations arranged in quincunx; triangular areas very indistinct; margin narrow, coriaceous, coated with a close, short down, alternately red and white.

Length 17.5, width 11.3 mill.

New England. (Eur.)

5. C. ALBUS, Montagu. Fig. 202.
    Test. Brit. 4. 1803.
Chiton asclloides, Lowe, Zool. Journ., ii. 103, t. 5, f. 3.
Chiton sagrinatus, Couthouy, Am. Journ. Sci., xxxiv. 217. 1838.

Shell small; valves with a small beak, minutely crenulate on their anterior margin, subcarinate, with minute striæ; surface minutely shagreened; an obsolete diagonal ridge sometimes divides each side into triangular areas, but for the most part without any distinct boundary; margin membranous covered with beaded granules; grayish-white under a pulverulent black epidermis; marginal membrane ashy, with a narrow black median band.

Length 10, width 3.7 mill.

New England. (Eur.)

6. C. EMERSONII, Couthouy.   Fig. 203.

    Am. Journ. Sci., xxxiv. 217.   1838.

Chiton restitus, Sowerby, Zool. Journ., iv. 368.

Shell ovate-oblong, broadest behind; valves uniform, each with a central heart-shaped area, with bead-like granules or tubercles in concentric series round the margin, the remainder covered with a soiled downy membrane; marginal membrane with series of yellow hairy tufts; whitish.

Length 20, width 12.5 mill.

New England.

7. C. RUBER, Lowe.   Fig. 204.

    Zool. Journ., iii. 101, t. 5, f. 2.

Shell small, oval, elevated, carinated; surface smooth under the lens, except the lines of growth; valves strongly beaked; light bright red or flesh-color under a blackish pigment; interior bright rose-red.

New England. (Eur.)

Distinguished from *C. marmoreus* by its unpunctured surface.

ORDER III. OPISTHOBRANCHIATA.

SECTION A. TECTIBRANCHIATA.   Animal usually provided with a shell, both in the larval and adult state; branchiæ covered by the shell or mantle; sexes united.

SECTION B. NUDIBRANCHIATA.   Animal destitute of a shell except in the embryo state; branchiæ always external, on the back or sides of the body; sexes united.

SECTION A.

Family TORNATELLIDÆ.   Shell external, solid, spiral or convoluted; subcylindrical; aperture long and narrow; columella plaited; sometimes operculated.

Family BULLIDÆ.   Shell invested by the animal, globular or cylindrical, convoluted, thin, often punctate striated; spire small or concealed; aperture long, rounded, and sinuated in front; lip sharp.   No operculum.

Family TORNATELLIDÆ.

Genus **TORNATELLA**, Lamarck.

Extr. d'un Cours.   1812.

Shell solid, ovate, with a conical, many-whorled spire; spirally grooved or punctate striate; aperture long, narrow, rounded in front; outer lip sharp; columella with a strong spiral fold; operculum horny, elliptical, lamellar.

Animal white: head truncated and slightly notched in front, furnished posteriorly with recumbent tentacular lobes, and small eyes near their inner bases; foot oblong, lateral lobes slightly reflected on the shell.

There are few living, but over seventy fossil species. Distribution universal.

1. T. PUNCTO-STIATA, Adams.   Fig. 205.

   Bost. Journ. Nat. Hist., 323, t. 3, f. 6.   1840.

Shell minute, suboval, polished; whorls four to five; body whorl large, smooth above the aperture; beneath it, with ten to fifteen punctate revolving lines; spire short, rapidly diminishing, with a shoulder near the suture; suture deeply impressed; aperture two-thirds the length of the body-whorl, becoming wider beneath; pillar lip with a prominent fold. Umbilicus open in young shells; white.

Length 2.5 to 3.7 mill.

New York to Massachusetts.

Genus **RINGICULA**, Deshayes.

Anim. sans Vert., viii. 341.   1838.

Shell small, ventricose, smooth or concentrically striated; spire small; aperture with an oblique notch in front; columella callous, strongly plicated; outer lip thickened and reflected, with a marginal callus.

1. R. NITIDA, Verrill.

   Am. Journ. Sci., 16th January, 1873.

Shell small, white, smooth, broad oval, with five whorls; spire rapidly tapering, subacute, shorter than the aperture; whorls very convex, with deep suture, and a subsutural impressed line; columella stout, recurved at the end, with two strong, very prominent equal, spiral folds, the anterior one projecting beyond the canal, with the end rounded.

Length 4 +, diam. 3 mill.

New England.

Family BULLIDÆ.

Synopsis of Genera.

Shell convolute, ovate or subglobose, smooth, generally mottled; spire involute sunken, causing the apex to be tubular or perforate; aperture extending the entire length of the body-whorl; inner lip simple; columella none; outer lip acute.   BULLA, Klein.

Shell solid, cylindrical, involute; spire none; apex obtuse, umbilicated; aperture narrow and linear, as long as the body-whorl; inner lip callous, with a single anterior fold; outer lip straight, simple.
       CYLICHNA, Lovén.

Shell rather thin, subcylindrical, imperforate, covered with an epidermis; spire distinct, apex obtuse, not mamillated, sutures simple, not canaliculated; aperture narrow behind, dilated and entire in front, nearly as long as the body-whorl; columella simple, not plicate; outer lip straight, acute.   UTRICULUS, Brown.

Shell thin, hyaline, subumbilicated, inflated, ovate or subglobose; spire depressed, with a mamillated nucleus; aperture expanded, not extending beyond the body-whorl; columella reflexed and sinuous; outer lip sinuous, produced anteriorly.   DIAPHANA, Brown.

Shell ovate-pyriform, convolute; spire distinct, depressed, somewhat concealed; aperture very wide, narrowed behind, entire and dilated in front; inner lip spirally convoluted as far as the commencement of the spire; outer lip simple, acute.   SCAPHANDER, Montfort.

Shell concealed in the mantle, loosely convolute, thin, fragile, suborbicular or ovate, striate or punctate; spire small, often concealed; aperture very wide and open; outer lip patulous.   PHILINE, Ascanias.

Genus BULLA, Klein.
Ostracol, 82. 1753.

The eyes are conspicuous, sessile on the middle of the frontal disk; mantle with the outer margin forming a thick fleshy lobe; foot with moderate lateral lobes partly investing the shell, the hind part not extending beyond the shell.

The species of this genus inhabit sandy mud-flats, the slimy banks of river-mouths, and brackish places near the sea. They feed on bivalves and other mollusks, which they swallow whole, reducing and crushing them afterwards by the calcareous or horny plates of their powerful, muscular gizzard. The shells are rather solid, smooth, or nearly so, and marbled and mottled like birds' eggs, or white.

There are about fifty species, inhabiting temperate and tropical seas, and ranging from low water to twenty-five fathoms.

1. B. INCINCTA, Mighels.
    Proc. Bost. Soc. Nat. Hist., i. 188. 1844.

Shell small, cylindrical, opaque, white; whorls three, the first slightly depressed, the last distinctly girded above the middle;

epidermis yellowish; spire obtuse, elevated; suture canaliculate; aperture narrow above, wide and rounded below; outer lip sharp, entire, advanced in the central region, with a fissure posteriorly. Length 3, diam. 1.5 mill.

Casco Bay, Me.

This species has not been figured, and I am not acquainted with it; nor has it been found since the date of the original description.

2. B. SOLITARIA, Say. Fig. 206.

Journ. Philad. Acad., ii. 245. 1822.

Bulla insculpta, Totten, Am. Journ. Sci., xxviii. 350, f. 4. 1835.

Shell small, thin, fragile, pellucid, oval, impressed at the top, with numerous microscopic revolving lines; spire none, but in its place a pit; aperture narrowly linear above, wide below; umbilicus none; white.

Length 9, diam. 6 mill.

Whole Coast.

3. B. OCCULTA, Mighels and Adams. Fig. 207.

Proc. Bost. Soc. Nat. Hist., i. 50. 1841.

Bulla Reinhardi, Möller, Ind. Moll. Grœn., 6. 1842.

Shell small, of a dingy-white color, ovate-cylindrical, covered with very minute transverse striæ, and with indistinct striæ of growth; spire concealed; labrum extends a little below the spire, nearly straight above the centre, regularly rounded below and at base; aperture narrow at the upper part, rather broad at the base. Length 5, diam. 3.75 mill.

New England to Greenland. (Eur.)

Mr. Jeffreys says this is identical with *Cylichna striata*, Brown, 1827.

Genus **CYLICHNA**, Lovén.

Ind. Moll. Scand. 10. 1846.

In this genus the tentacular lobes are connate, indistinct; eyes sessile on their front bases; mantle with a thick posterior lobe, partially closing the aperture of the shell.

The species chiefly inhabit deep water, and the genus is of world-wide distribution.

1. C. ALBA, Brown. Fig. 208.

(*Volvaria.*) Brit. Conch., 3, t. 88, f. 43-44. 1827.

Bulla triticea, Couthouy, Bost. Journ. Nat. Hist., ii. 88, t. 2, f. 8. 1838.

Bulla corticata, Möller, Ind. Moll. Grœn., 6. 1842.

Shell polished, cylindrical, rather solid; spire slightly depressed, imperforate; surface reticulated by fine microscopic striæ; lip arising from the margin of the circular pit at the summit of the spire; aperture linear above, broad below; umbilicus covered with enamel, which gradually disappears within the aperture; white under a ferruginous epidermis.

Length 7.5, diam. 2.5 mill.

New England, northwards. (*Eur.*)

2. C. ORYZA, Totten. Fig. 209.

(*Bulla.*) Am. Journ. Sci., xxviii. 350, t. 5. 1835.

Shell minute, not very thin; tip depressed into a shallow pit, and base rather acute; aperture as long as the shell, narrow above and wider below; outer lip sharp, regularly arched rising a little higher than the shoulder; an oblique, truncated fold on the columella; white.

Length 7.5, diam. 2.5 mill.

New England, south of Cape Cod. (*Eur.*)

According to Mr. Jeffreys this shell = *B. utriculus*, Brocchi, 1814.

Genus UTRICULUS, Brown.
Brit. Conch. 1844.

Head disk very short; tentacular lobes lateral, rounded; eyes none.

1. U. GOULDII, Couthouy. Fig. 210, 211.

(*Bulla.*) Bost. Journ. Nat. Hist., ii. 181, t. 4, f. 8. 1839.

U. turritus, Möller, Ind. Moll. Groenl. 1842.

Bulla pertenuis, Mighels, Proc. Bost. Soc., i. 129. 1843.

Shell thin, small, of four whorls, rounded at their upper edges, with well-defined sutures and fine transverse striæ; spire depressed, discoidal, sometimes slightly mammillated; no umbilicus; white, with yellowish epidermis.

Length 7.5, diam. 2.5 mill.

New England. (*Eur.*)

The *Bulla pertenuis* of Mighels (Fig. 211) appears to be the young of *Gouldii*.

2. U. CANALICULATUS, Say. Fig. 212.

(*Volvaria.*) Journ. Philad. Acad., v. 211. 1826.

Bulla obstricta, Gould, Am. Journ. Sci., xxxviii. 196. 1840.

Shell minute, cylindrical, polished, with very faint lines of growth; spire convex, a little elevated, with minute but prominent tip; whorls about five, with their shoulders very obtusely grooved; outer lip arching forward; inner lip with a thin coat of enamel, with a single oblique fold or small tooth near the base; whitish, immaculate.

Length 2.5 to 5 mill.

New England; South Carolina.

3. U. BIPLICATUS, H. C. Lea. Fig. 213.

(*Bulla.*) Proc. Bost. Soc. Nat. Hist., 204. 1844.

Shell cylindrical, subquadrate, thick, whitish, polished, ivory-like; spire concealed; last whorl with a callus above, and small transverse striæ below; mouth narrow above, ovate below; columella with a large and a small fold.

Cape May, N. J.

Genus DIAPHANA, Brown.
Conch. Text-Book, 112. 1833.

Head-disk broad and short; tentacular lobes short, conical, lateral, wide apart; eyes immersed in their bases behind; mantle-margin slightly thickened; foot short, bilobed behind.

There are but few species, of northern distribution.

1. D. HIEMALIS, Couthouy. Fig. 214.

(*Bulla.*) Bost. Journ. Nat. Hist., ii. 180, t. 4, f. 5. 1839.

Utriculus globosus, Lovén, Ind. Moll. Scand. 1846.

Shell globular, very thin and brittle; the body-whorl enveloping all the others so as to leave no perceptible spire, and marked with the lines of growth; aperture narrow above, dilated beneath; outer lip strong, and regularly curved; it revolves from its junction behind nearly a third of a revolution before it turns forward; columella slightly arcuated and reflected upon the body of the shell so as to form a small but distinct umbilicus; hyaline, with brownish tinge.

Length 2.5. mill.

New England, northwards. (*Eur.*)

2. D. HYALINUS, Turton. Fig. 215.

(*Utriculus.*) Loudon's Mag. Nat. Hist., vii. 353. 1833.

Bulla debilis, Gould, Am. Journ. Sci., xxxviii. 196. 1840.

Shell small, obliquely ovate, tumid. thin and brittle; whorls four, all rising to the same height, convexly rounded; last whorl the whole length of the shell; surface smooth; aperture as long as the shell, widening below, outer lip slightly waved, inner lip spread out into a thin enamel upon the body of the shell, partially covering an umbilical indentation; greenish-white.

Length 2.5 mill.

New England, northwards. (*Eur.*)

Genus SCAPHANDER, Montfort.

Conch. Syst., ii. 335. 1810.

Animal not investing the shell; eyes none; foot ample, but short, the side lobes small.

1. S. PUNCTO-STRIATUS, Mighels and Adams. Fig. 216.

(*Bulla.*) Proc. Bost. Soc. Nat. Hist., i. 49. 1841.

S. librarius, Lovén, Ind. Moll. Scand. 1846.

Shell white, rather solid, ovate, with crowded inequidistant punctate striæ; spire concealed; aperture very large, labrum rising above the apex, very sharp and regularly arcuate: labium with a very thin lamina extending to the apex.

Length 10, diam. 6 mill.

Casco Bay, Me. (*Eur.*)

Mr. A. E. Verrill has recently obtained specimens in deep water off St. George's Bank, measuring over an inch in length.

Genus PHILINE, Ascanias.

Act. Holm. 1772.

Animal investing the shell; eyes none; foot not produced posteriorly, the side lobes large and fleshy; shell concealed in the mantle.

The animals composing this genus are blind, like most creatures that seek their food by burrowing. They frequent mud-flats and slimy banks at the entrances of rivers, which they perforate near the surface, and probe with their flattened heads for the small bivalves which constitute their prey; these they seize and swallow entire, breaking their shells by means of their testaceous, muscular gizzards. There are about twenty species; distribution universal.

1. P. SINUATA, Stimpson. Fig. 217.

Proc. Bost. Soc. Nat. Hist., iii. 333. 1850.

Shell minute, ovate, white, pellucid, longitudinally striate; spire conspicuous; aperture anteriorly dilated.

Length 1.75, diam. 1.25 mill.

The animal is yellowish, elongated; darkest behind, with dots and patches of white.

The ova are deposited in the latter part of August. They are minute, white, and enveloped in a gelatinous mass, which is globular, hyaline, and somewhat larger than the animal.

Massachusetts.

2. P. QUADRATA, Searles-Wood.   Fig. 218.

Mag. Nat. Hist. n. ser., iii. 461, t. 7, f. 1.   1839.

Philine formosa, Stimpson, Bost. Proc., iii. 334.   1850.

Shell minute, squarely globose, subtruncated anteriorly, white, shining, subopaque, thickened posteriorly, punctured with inequidistant, sometimes undulated, revolving striæ; apex deeply excavated, columella sinuose, broadly and lightly callous; lip crenulated posteriorly; aperture very wide.

Length 4.5, diam. 3.5 mill.

New England, northwards.   (*Zetlands.*)

3. P. LINEOLATA, Conthouy.   Fig. 219.

(*Bulla.*)   Bost. Journ. Nat. Hist., iii. 179, t. 3, f. 15.   1832.

Shell very small, oblong-ovate, broadest at the base, thin and fragile; whorls three; the last inflated and enveloping all the others, with numerous impressed minute revolving striæ; spire small, prominent, flattened, with the outer lip arising from near its summit; aperture the whole length of the shell, narrow above, dilated beneath, somewhat effuse at the base; a faint oblique fold near the middle of the columella; pale-brown, with a thin ferruginous epidermis; within glossy yellowish-white.

Length 3.75 mill., diam. 2 mill.

Massachusetts, northwards.

According to Mr. Gwyn Jeffreys this species = *lima*, Brown, 1827.

Section B. NUDIBRANCHIATA.

Gills exposed or contractile into cavities on the surface of the mantle; adult animal without any shell; larva shell-bearing; foot elongate, formed for walking; sexes united.

* *Gills plumose, on the hinder part of the mantle, disposed in a circle or semicircle, round the vent. Anthobranchiata.*

Family DORIDIDÆ.   Mantle-edge simple; gills surrounding the vent, on the middle of the hinder part of the back, in a common cavity.

The gills are retractile into a common cavity, and the mantle is very large, either entirely or almost covering and concealing the foot.

Family TRIOPIDÆ. Mantle small, edged with tentacular appendages; gills on the middle of the hinder part of the back, in a common cavity surrounding the vent; vent dorsal.

* *Gills various, not arranged round the vent but usually in rows along the sides of the body.*

Family TRITONIIDÆ. Tongue broad, teeth many in each cross series; jaws horny; gills superficial, fusiform, or branched on each side of the back; vent lateral; foot linear, channeled.

Family DOTONIDÆ. Tongue narrow; teeth in a single central series; tentacles sheathed at the base, retractile; gills superficial, fusiform, on the sides of the back.

Family ÆOLIDIDÆ. Tongue narrow; teeth in a single central series; jaws horny; tentacles subulate, simple, rarely ringed, contractile; gills superficial, fusiform, or branched on the sides of the back; vent lateral.

Family HERMÆIDÆ. Body elongated, not provided with a distinct mantle; mouth unarmed, or with corneous jaws; tentacles sometimes wanting; when present two, dorsal, non-retractile; gills papillose; vent usually central, on the posterior half of the back; genital orifice at the right side.

The dorsal position of the vent and the indistinct mantle distinguish this family from *Æolididæ*, and the presence of papillose gills from *Elysiidæ*.

Family ELYSIIDÆ. Body limaciform, clothed with cilia; tongue narrow; teeth in a single central series; tentacles subulate or linear, folded; eyes sessile, near the bases of the tentacles; gills in the form of plaits or vessels radiating on the surface of the back; vent central, dorsal on the hinder part of the back.

Family LIMAPONTIIDÆ. Body depressed; tongue narrow; teeth in a single, central series; tentacles none, or simple, contractile; gills none external.

Family DORIDIDÆ.

These animals are most attractive nudibranchs, and may be studied by placing them in glass reservoirs of salt-water, as they are by no means shy, but extend their tentacles and display their branchial plumes to great advantage. In this family the gills are retractile into a common cavity, and the mantle is very large, either entirely or almost covering and concealing the foot.

Synopsis of Genera.

Subfamily DORIDINÆ.

Body depressed, rounded above ; mantle convex, large, simple, covering the head and foot.

Tentacles dorsal, subclavate, laminated, retractile within a cavity ; gills arborescent, retractile ; vent in the centre of the gills.

DORIS, Linnæus.

Body covered with an ample, smooth mantle, oval, convex ; dorsal tentacles retractile, without sheaths ; head prominent, the lateral angles prolonged anteriorly as short oval palpi or tentacles ; foot broad, cordate ; branchiæ posterior, in the groove between the mantle and foot. DORIDELLA, Verrill.

Subfamily POLYCERINÆ.

Body elongate, subangular ; mantle indistinct.

Body smooth or tuberculated ; tentacles clavate, pectinate, non-retractile, without sheaths ; a frontal veil with simple processes on the head ; gills with two or more lateral appendages.

POLYCERA, Cuvier.

Genus **DORIS**, Linnæus.
Syst. Nat., edit. x. 653. 1758.

The branchial plumes form an elaborate coronal around the vent, which, viewed with a common lens in a vessel of water, forms, when fully expanded, a beautiful object. The surface of the mantle is either smooth or tubercular, and the sheaths of the tentacles are often crenate on their margins.

1. D. BILAMELLATA, Linnæus. Fig. 220.
Syst. Nat., edit. xii. 1083. 1767.
Doris fusca, Muller. Zool. Dan. Prodr. 229. 1780.
Doris verrucosa, Pennant. Brit. Zool., iv. 43, t. 21, f. 23. 1777.
Doris vulgaris, Leach, Syn. Moll. Gr. Brit. 19.
Doris Elfortiana, Leach, Ibid., 20, t. 7, f. 1.
Doris affinis, Thompson. Ann. Nat. Hist., v. 85.
Doris liturata, Beck, Möller, Ind. Moll. Grœnl., 5.
Doris obrelata, Bouchard, Cat. Moll. Boul., 42.
Doris coronata, Agassiz, Proc. Bost. Soc. Nat. Hist., iii. 191.

Body elliptical, covered with pestle-shaped papillæ, whitish varied with rusty brown or flesh-color, and opaque white ; branchiæ twenty to twenty-five, long, linear, simply pinnate, arranged transversely in an oval, including several tubercles.

Length about an inch, width half an inch.

New England to Greenland, N. Europe.

2. D. TENELLA Agassiz. Fig. 221.

Gould. Invert. Mass., 2d edit. 229, t. 20, f. 289, 290, 293. 1870.

Body ovate, covered with small, pointed tubercles, yellowish-white; branchial plumes six to seven, simple; mantle extended anteriorly beyond the foot, head dilated laterally.

Length half an inch, breadth three-eighths of an inch.

Massachusetts.

This and the following species are both referred by Verrill (*Am. Journ. Sci.*, ii. 407, 1870) to the genus *Onchidoris*, which differs from *Doris* partly in the gills being retractile into separate cavities.

3. D. ASPERA, Alder and Hancock. Fig. 222.

Ann. Nat. Hist., ix. 32.

D. pallida, Agassiz, Proc. Bost. Nat. Hist., iii. 191. 1849.

Proctaporia fusca, Mörch. Greenl. Blöddyr, 6. 1857.

Body elongated, sides parallel, ends equally rounded, covered with large mushroom-like tubercles, cream-colored; branchial plumes seven to eight, simple, retractile.

Length half an inch, breadth one-fourth of an inch.

Gould calls this species *D. pallida*, Agassiz, and writes (Invert. Mass, edit. ii. 230), " It is pretty certainly *D. aspera*, Alder and Hancock, but if the exhibition of a colored drawing is a valid claim, the name of Agassiz has precedence." If the drawing had been *published* with a name attached to it, it would have secured priority, but as it was merely shown at a meeting of the Boston Society, the claim is simply ridiculous.

Massachusetts. (*Eur.*)

4. D. TUBERCULATA, Cuvier, 180, 2, 1802. Fig. 223.

D. diadematа, Agassiz, Gould, Invert. Mass., 2d edit. 230, t. 21, f. 298, 300 to 304. 1870.

Body oblong-oval, slightly broader anteriorly, maroon-color, darkened on the sides by numerous dusky points, dark gray below; branchial plumes nine, simple; head short, concentric, pointed.

Length one and a half inch, breadth nearly an inch.

Massachusetts.

5. D. REPANDA, Alder and Hancock. Fig. 224.

Ann. Nat. Hist., ix. 32.

Doris planulata, Stimpson, Invert. Grand Manan, 26, f. 14. 1853.

Body broad, mantle extended beyond the foot, covered with

white minute tubercles; white, with a row of irregular bright-yellow spots down each side; branchial plumes ten, small, pinnated.

Length 15, breadth 12 mill.

New England, N. Europe.

6. D. GRISEA, Stimpson. Fig. 225.

Gould, Invert. Mass., 232, t. 20, f. 292, 295. 1870.

Body oblong-oval, covered with blunt processes tipped with stellate clusters of spiculæ; branchial plumes short, yellowish, arranged in a circle around a dark bristle; head short, broad, angular.

Length 13, breadth 9 mill.

Massachusetts.

Verrill (*Am. Journ. Sci.*, l. 408, 1870) places this species in the genus *Onchidoris*.

7. D. BIFIDA, Verrill.

Am. Journ. Sci., l. 406. 1870.

Broadly oval, widest anteriorly, back very convex, mantle covered with pointed papillæ. Tentacles rather long, thickest in middle, the outer half strongly plicated, but with a smooth tip, the base surrounded by small papillæ. Dark purple-brown, with white spots; the edges and tips of the gills yellow.

Length 1 inch, breadth ½ inch.

Eastport, Maine.

Genus DORIDELLA, Verrill.
Am. Journ. Sci., l. 408. 1870.

1. D. OBSCURA, Verrill. Fig. 226.

Am. Journ. Sci., l. 408, f. 2. 1870.

Broadly oval; back convex, smooth. Foot broad, cordate in front. Oval disk broad, emarginate or with concave outline in front; the angles somewhat produced, forming short, tentacle-like organs. Dorsal tentacles small, stout, retractile. Color blackish, lighter towards the edge, as if covered with nearly confluent black spots; foot, oral disk, and dorsal tentacles white; the central part of the body, beneath, bright yellow.

Length 7.5, breadth 5 mill.

New Haven, Conn.

Genus POLYCERA, Cuvier.
Regn. Anim. II. 390, 1817.

Animal smooth or tuberculated; tentacles clubbed and pectinated, not retractile, and without sheaths; frontal veil consisting
of a series of tentaculiform appendages variable in number, often
extending along the borders of the mantle; branchiæ forming
part of a circle around the vent, encased by membranous laminæ
which protect them.

1. P. LESSONII, Orb. Fig. 227.
  Mag. de Zool., vii. 5, t. 105.
  Polycera citrina, Alder, Ann. Nat. Hist., vi. 304, t. 9, f. 1–9. 1841.
  Polycera modesta, Lovén, Index Moll. Grœnl., 6. 1846.
  Doris illuminata, Gould, Invert. Mass., edit. i. 4. 1841.
  Massachusetts; North Europe.

Family TRIOPIDÆ.

In this family the body is somewhat angular, and the mantle is
distinct and furnished with tubercular appendages; the species
of the genera comprising this group constitute some of the most
delicate and beautiful forms of nudibranchiate mollusks.

Genus ANCULA, Lovén.
Ind. Moll. Scand., 5. 1846.

Body slender, with clavate processes bordering the branchial
region of the back; tentacles clavate, perfoliate, laminated, armed
at the base with styliform appendages.

1. A SULPHUREA, Stimpson. Fig. 228.
  Invert. Grand Manan, 26. 1853.

Body long, slender, light-brownish; branchial plumes three,
arranged in a semicircle, anterior largest, doubly pinnate; surrounding tentacular processes eight to twelve, sulphur-tipped;
oral tentacles long, the processes arising from their very base.
Length 30 mill.

Family TRITONIIDÆ.

Many of the genera of this family are pelagic, and are often
found crawling on the fronds of floating algæ or clinging to the
narrow stems of gulf-weed, which is frequently met with in large
masses at considerable distance from the land; these mimic forests-

tenanted by their singular molluscan inhabitants, thus serve in some measure to enliven the solitude of the ocean.

Genus DENDRONOTUS, Alder and Hancock.

Tentacles clavate, laminated; front of head with branched appendages; gills ramose, arranged in a single series down each side of the back.

1. D. ARBORESCENS, Müller. 229.

(*Doris.*) Zool. Dan. Prodr., 229. 1780.

Tritonia Reynoldsii, Couthouy, Bost. Journ. Nat. Hist., ii. 74, t. 2, f. 1–4. 1838.

Tritonia lactea, Thompson, Ann. Nat. Hist., v. 88, t. 2, f. 3.

Tritonia pulchella, Alder and Hancock, Ann. Nat. Hist., ix. 33.

Body tapering to the tail, which ends acutely; sides with numerous papillæ; head short, depressed, orbicular, supporting three pair of gills; mouth crescent-shaped, papillose, with strong transverse folds; jaws angular; tentacula arising from the back of the head, and received into a round sheath, which terminates in five unequal branches; five pair of dorsal gills, all susceptible of being retracted into the body of the animal, leaving in their places small tubercles; rufous brown, occasionally dark brown, with patches of white on the back between the branchial tufts; foot white, diaphanous.

Length 3.5 inches.

New England; Northern Europe.

2. D. ROBUSTUS, Verrill.   Fig. 230.

Am. Journ. Sci., l. 405, f. 1.   1870.

Eastport, Maine; Grand Manan Island. (*Eur.*)

Family DOTONIDÆ.

Genus DOTO, Oken.

Lehrb. Naturg. 1815.

Head covered by a simple veil; tentacles linear, sheaths trumpet-shaped; gills clavate, compound, or rough, with whorls of tubercles ranged in a single series on each side of the back.

The tentacular sheaths have simple margins, and the ovate branchiæ are rough with tubercles; the front of the head is simple, and the foot is linear; they appear to feed on hydroid zoophytes.

1. D. CORONATA, Gmelin.   Fig. 231.
    (*Doris.*) Syst. Nat. i. 3105. 1790.
Melibœa coronata, Johnston, Ann. Nat. Hist., i. 117, t. 3, f. 5–8.

Animal yellowish, dotted with red; veil square in front; branchiæ five to seven on each side, ovate club-shaped, bearing several circles of papillæ with dark red tips.
    Length half an inch.

New England. (*Eur.*)

Family ÆOLIDIDÆ.

Synopsis of Genera.

Body broad; tentacles four, smooth, elongate, subulate; labial feelers elongate; gills papillose, arranged in longitudinal rows, not clustered, numerous, depressed, and imbricated.   ÆOLIS, Cuvier.

Tentacles subulate, annulate, or perfoliate; labial feelers subulate; gills clustered, or arranged in separate tufts along the back.

FLABELLINA, Cuvier.

Body linear: tentacles subulate, smooth, simple; labial feelers short; gills in a single row on each side; foot square in front.   TERGIPES, Cuvier.

Head without tentacles; labial feelers very long and tapering; gills pyriform, placed in longitudinal lines; front of foot angular.

CALLIOPÆA, D'Orbigny.

Genus ÆOLIS, Cuvier.
Tabl. Elem. 1798.

1. Æ. PAPILLOSA, Linnæus.   Fig. 232.
    (*Limax.*) Syst. Nat., edit. xii. 1082. 1767.
Eolis farinacea, Fould, Stimpson, Invert. Grand Manan, 25. 1853.[1]

Animal ovate-oblong, depressed, dusky or orange-colored, dotted with brown, ochreous, or white; branchiæ numerous, somewhat compressed, crowded and imbricated, eighteen to twenty-four oblique ranges; dorsal tentacles short, smooth, conical, labial tentacles short and simple; angles of foot slightly prolonged.
    Length two to three inches, breadth one-third the length.

New England; Northern Europe.

2. Æ. SALMONACEA, Couthouy.   Fig. 233.
    (*Eolis.*) Bost. Journ. Nat. Hist., ii. 68, t. 1, f. 2. 1838.
Eolis Bodoensis, Möll. Moll. Græn. 1842.

Body nearly diaphanous; back with a conspicuous elevation in the middle; head large, with four tentacula; the superior minutely

---

[1] There are about a dozen additional synonymic names by British authors.

8

serrated; mouth Λ-shaped; branchiæ in longitudinal series, one hundred or more; foot with two short processes in front, and ending in a point behind; pale yellowish-white; branchial cirri salmon-colored, bordering on orange.

Length one and three-quarters inch.

Charles River, Mass.

Genus FLABELLINA, Cuvier.

1. F. BOSTONIENSIS, Couthouy.   Fig. 234.

(*Eolis.*)   Bost. Journ. Nat. Hist., ii. 67, t. 1, f. 1.   1838.

Body elongate, lanceolate, delicate drab-color, with a silvery line on the tail and on the back of the anterior tentacles, which are long, subulate; posterior tentacles shorter, serrated at tips; branchiæ curved lanceolate, nucleus drab-colored, tips white, in four to six distant groups on each side; angles of foot much produced.

Length 25, diam. 7.5 mill.

Massachusetts.

2. F. RUFIBRANCHIALIS, Johnston.   Fig. 235.

(*Eolidia.*)   Lond. Mag. Nat. Hist., v. 428.

Eolidia Embletoni, Johnston, Ibid., viii. 79.

Eolis Mananensis, Stimpson, Invert. Grand Manan, 26.   1853.

Body slender, tapering, white; oval and dorsal tentacles subequal; branchiæ nearly linear, variable in length, disposed in six or seven clusters on each side, interior of a bright vermillion, with an opaque white rim near tip; anterior angles of foot prolonged and folded transversely.

Length one inch.

New England, northwards.   (*Eur.*)

3. F. PILATA, Gould.   Fig. 236.

Invert. Mass., 2d edit. 243, t. 19, f. 270, 277, 279, 281.   1870.

Body elongated, a carmine line margined with silvery dots between the tentacles and each tuft of branchiæ, tail silvery; tentacles subulate, simple, tipped with silvery, branchiæ clavate, contracted at tip, which has two silvery zones, nucleus pale-chestnut, arranged in five or more distant groups of two transverse ranges.

Length 37, breadth 6 mill.

Charles River, Mass.

4. F. STELLATA, Stimpson.   Fig. 237.

(*Eolis.*)   Invert. Grand Manan, 25.   1853.

Body slender, pale-white; dorsal tentacles wrinkled transversely, long, but shorter than oral; branchiæ few, arranged in about five clusters on each side, those of second and third being longest, giving a star-like appearance to the animal when rolled up; foot strongly auricled in front.

Length 10 mill.

Grand Manan Island.

5. F. PURPUREA, Stimpson.
   (*Eolis.*) Invert. Grand Manan, 25. 1853.

Body large, full, robust, tentacles rather short, thick, smooth; the dorsal ones with the eyes far behind their bases; papillæ large, flattened, crowded, arranged in five or six clusters on each side, leaving the middle third of the body bare; foot broad, with short auricles in front; mouth disk large, triangular; body pale-whitish, dark in the middle line, from the viscera showing through; papillæ dark-purplish, with the tips covered with intense white specks.

Length 1 inch.

Duck Island.

6. F. PICTA, Alder and Hancock.   Fig. 238.
   (*Eolis.*) Monog. Nud. Moll., t. 33.   1847.

Yellowish-white, blotched with brownish-amber; oral tentacles short, stout; dorsal tentacles twice as long, simple, with an amber ring at outer third; branchiæ like an olive-jar, arranged in six or eight series; foot narrower than body, obtuse posteriorly, anterior angles rounded.

Length 18, breadth 4.5 mill.

Massachusetts.

7. F. DIVERSA, Couthouy.   Fig. 239.
   (*Eolis.*) Bost. Journ. Nat. Hist., ii. 187, t. 4, f. 14.   1839.

Body lanceolate, acutely pointed, pale-yellow; oral tentacles long and delicate; dorsal tentacles shorter, linear; branchiæ lanceolate, externally transparent and colorless, interior orange, thickly arranged along the sides in transverse series of three or four; foot with the angles slightly dilated.

Length 31, breadth 8 mill.

New England; Grand Manan.

Genus **TERGIPES**, Cuvier.

Ann. du Mus., xix. 1812.

1. T. DESPECTA, Johnston.   Fig. 240.

(*Eolis.*) Mag. Nat. Hist., viii. 378, f. 35*e.*

Animal colorless, with a zigzag olive stripe along the back; branchiæ large, ovate, in a single series along each side; dorsal tentacles long; angles of foot not produced.

Length 6, breadth 1.25 mill.

Massachusetts; Scotland.

2. T. GYMNOTA, Couthony.   Fig. 241.

(*Eolis.*) Bost. Journ. Nat. Hist., ii. 69, t. 1, f. 3.   1838.

Animal small, tapering to a fine point, watery white; tentacles short, the posterior pair minutely serrated; branchiæ in seven lateral clusters of about five each, slightly club-shaped, having a reddish-brown centre.

Length 25, diam. 2.5 mill.

Massachusetts.

Genus **CALLIOPÆA**, Orb.

Mag. Zool., t. 108.   1837.

1. C. FUSCATA, Gould.   Fig. 241*a.*

Invert. Mass., edit. ii. 250, t. 16, f. 218–221.   1870.

Animal semicylindrical, attenuated behind, dark slate-colored; head not distinct, excavated in front; tentacles two, long, pointed; branchiæ long, club-shaped, slender at base, alternating in two parallel rows on the two posterior thirds of each side, the lower series much the smaller; foot bilobed in front, contracted posteriorly.

Length 7.5, diam. .8 mill.

Massachusetts.

Family HERMÆIDÆ.

Synopsis of Genera.

Tentacles two, longitudinally folded; head without lobes; gills elongate, papillose, smooth, arranged along the sides of the back.

HERMÆA, Lovén.

Head without tentacles, produced into a lobe on each side; gills papillose, arranged in transverse rows on the sides of the back.

ALDERIA, Allman.

Tentacles two, simple, undefended, contractile, sublateral; a large labial veil, produced on each side into an oblong, flat lobe; gills simple, papillose, lateral, in a simple series on each side of the back; foot rather broad.

CLŒLIA, Lovén.

Genus **HERMÆA**, Lovén.

Ofvers. Kong. Vet. Acad. Handl. 1844.

1. H. CRUCIATA, Alex. Agassiz. Fig. 242.

Gould, Invert. Mass., edit. ii. 253, t. 17, f. 256. 1870.

Body very slender, the tail much attenuated; foot narrower than the body, obtusely dilated at the anterior angles; head small, semicircular; mouth inferior; tentacles dilated and obtusely pointed, the superior face longer than the inferior; branchiæ dilated, shaped much like trefoil or the ace of clubs, the biliary organs within having a rude cruciate form; there are seven principal ones on each side, and eight or ten intermediate much smaller ones.

Massachusetts.

Genus **ALDERIA**, Allman.

Thompson, Rep. Faun. Ireland. 1844.

1. A. HARVARDIENSIS, Agassiz. Fig. 243.

Gould, Invert. Mass., edit. ii. 254, t. 16, f. 226–228. 1870.

Animal broad lanceolate, ochreous-brown; foot yellow; lateral prolongations of head tentacular; branchiæ short, curved, enlarging towards the tips, in about six clusters, of two each, on either side, of which the lower one is much smaller.

Length 12, breadth 4 mill.

Massachusetts; Grand Manan.

Genus **CLŒLIA**, Lovén.

Embletonia, Alder and Hancock, Ann. Mag., viii. 294. 1851.

1. C. FUSCATA, Gould. Fig. 244.

(*Embletonia.*) Invert. Mass., edit. ii. 251, t. 16, f. 229–232. 1870.

Animal subcylindrical, narrowing backwards; tail short, pointed, drab-colored; head larger than body, broad angles rounded, slightly emarginate; tentacles short; branchiæ club-shaped, arranged in five or six tufts on each side of the posterior portion of the body; angles of foot not dilated.

Length 4, breadth .66 mill.

Massachusetts.

2. C. BEMIGATA, Gould. Fig. 245.

(*Embletonia.*) Invert. Mass., edit. ii. 252, t. 16, f. 214–217. 1870.

Animal long, slender, uniform pale-yellow; head large, emarginate, angles dilated into triangular lobes with blunt points; tentacles long, linear; branchiæ removed from head, arranged on each side in distant tufts, the last pair at extremity of tail.

Length 6, breadth 1.2 mill.

Massachusetts.

Family ELYSIIDÆ.

In this family the respiratory function appears to be performed by the entire surface of the body, special organs for that purpose being almost obsolete.

Synopsis of Genera.

Body with the lateral ridges dilated into wing-like natatory appendages, not united ; head distinct, with two conspicuous auriform tentacles.

ELYSIA, Risso.

Body with lateral lobes united together posteriorly over the back.

ELYSIELLA, Verrill.

Genus ELYSIA, Risso.

Journ. Phys., 376. 1818.

1. E. CHLOROTICA, Agassiz.   Fig. 246.

Gould, Invert. Mass., edit. ii. 255, t. 17, f. 251–255. 1870.

Animal emerald-green, dotted with white and red spots; slender, tapering behind, with broad lateral expansions, folded and overlapping each other on the back when the animal is in motion; tentacles two, lanceolate, folded beneath; head distinct, obtuse, slightly emarginate; anterior angles of foot widely produced, triangular.

Massachusetts.

Genus ELYSIELLA, Verrill.

Am. Journ. Sci., ser. III., iii. 283, t. 7, f. 5.   1872.

1. E. CATULUS, Agassiz.   Fig. 247.

(*Placobranchus.*) Gould, Invert. Mass., edit. ii. 256, t. 17, f. 249, 250. 1870.

? *Placobranchus simplex*, Girard, Bost. Proc., v. 89. 1854.

Animal sea-green with whitish spots; body ovate-lanceolate; lateral expansions two-thirds its length, not meeting when reflected over the back; head large, rounded, globose; tentacles short, blunt, broad; foot wide as body, square in front, pointed behind.

Length 6, breadth 2.3 mill.

Massachusetts.

Family LIMAPONTIIDÆ.

In this group of slug-like forms the branchial appendages are altogether absent, or represented only by simple lobes or ridges on the sides of the body; the tentacles are linear, and not longitudinally folded as in Elysiidæ, and the body is depressed.

Genus LIMAPONTIA, Forbes.
London's Mag. Nat. Hist., v. 979. 1832.

Head elevated at the sides into two crest-like ridges; eyes large, sessile on the back of the head, in the centre of pale circular spaces; mantle distinct.

Gregarious, feeding on confervæ in small pools above half-tide.

1. L. ZONATA, Girard.
    (*Niobe.*)   Proc. Bost. Soc. Nat. Hist., iv. 211. 1852.

Less than a line in length; separation of body and head not distinctly defined; pale-reddish with transverse white bands.

Boston Harbor, Mass.

ORDER IV. **PTEROPODA**.

* *Head indistinct, with two wings on the sides of the mouth; gills internal;* body inclosed in a shell.   Thecosomata.

Family CAVOLINIDÆ. Animal with two united fins without any posterior foot-like appendage between them; abdomen voluminous; gills in pairs; shell calcareous, symmetrical, elongate or globular.

Family CYMBULIIDÆ. Animal globular or ovate; fins two, horizontal, opposite, on each side of the mouth, with a small intermediate lobe; shell cartilaginous, slipper-shaped, rarely wanting.

Family LIMACINIDÆ. Animal elongate, spiral; mouth at the union of the two fins and intermediate lobe, with two small labial swellings; fins elongate, rounded and united at their base by an intermediate lobe bearing an operculum; mantle large, open in front, forming a large gill cavity; gills internal; shell spiral, transparent; operculum distinct, spiral, vitreous, of few whorls.

** *Head distinct; wings two, or four, at the junction between the head and the body, with a central intermediate lobe or rudimentary foot; gills exterior; no shell.*   Gymnosomata.

Family CLIONIDÆ. Animal fusiform; head with a series of conical prominences on each side; wings two.

Family CAVOLINIDÆ.

Synopsis of Genera.

Body short, sometimes furnished with lateral appendages; shell globular; mouth narrower than the internal cavity, with a lateral slit on each side, interrupted in front.   CAVOLINA, Gioëni.

Body short, sometimes furnished with lateral appendages; shell globular, mouth narrower than the cavity, with a slit on each side, not interrupted in front; apex often truncated in the adult.   DIACRIA, Gray.

Animal elongate, conical, without lateral appendages; shell elongate, angular, conical; mouth larger than the cavity, without any lateral slits.   CLIO, Browne.

Body elongate, conical, rounded; shell elongate, conical, subcylindrical; mouth larger than the cavity, without any lateral slit.   STYLIOLA, Lesueur.

Genus CAVOLINA, Gioëni.
Desc. 1783.

1. C. TRIDENTATA, Gmelin.   Fig. 248.
    Syst. Nat., 3348. 1790.

Shell yellowish, pellucid, thin, very finely striated transversely; terminal tooth longer than the lateral ones.
Martha's Vineyard.
Dredged 20–25 fms. The distribution of this species is universal.

Genus DIACRIA, Gray.
Syn. Brit. Mus. 1840.

1. C. TRISPINOSA, Lesueur.   Fig. 249.
    Blainv. Dict. Sc. Nat., xxii. 82. 1782.

Shell elongated, straight, dilated anteriorly, compressed on each side, terminated posteriorly, with a very long spine, armed laterally with two short spines.
Nantucket (Europe, Rio Janeiro).

Genus CLIO, Browne.
Hist. Jamaica, 386. 1756.
Cleodora, Peron et Lesueur, Ann. Mus., xv. 1810.

1. C. PYRAMIDATA, Linnæus.   Fig. 250.
    Gmelin, Syst. Nat., 3148. 1790.

Shell triangular, pyramidal, short; mouth obliquely truncated.

Blainville, Man. Malacol. 1825.
Cresseis, Rang. Ann. Sci. Nat., xiii. 302. 1828.

1. S. VITREA, Verrill. Fig. 251.
Am. Journ. Sci., 3d ser. III. 248, t. 6, f. 7. 1872.

Shell smooth, polished, diaphanous, almost glassy, long conical, rather slender, slightly curved towards the acute apex.

Length 12, diam. 2 mill.

Animal white; swimming organs obovate, with the end broadly rounded, and bearing the slender tapering tentacles near the middle of the anterior edge; intermediate lobe short, rounded in front.

Martha's Vineyard.

Family CYMBULIIDÆ, Gray.
Syn. Brit. Mus. 1840.

Genus **PSYCHE**, Rang.
Ann. Sci. Nat., v. 284. 1825.

1. P. GLOBULOSA, Rang. Fig. 252.
Ann. Sci. Nat., v. 263. 1825.

Body round, diaphanous, mouth slightly arched, fins long, rounded at their extremity, narrowed at their base, with a slight shell-case above; viscera of a handsome purple, forming an ovoid mass, suspended in the middle of the body.

St. Pierre and Miquelon.

Family LIMACINIDÆ, Gray.
Syn. Brit. Mus. 1840.

Synopsis of Genera.

Shell subglobose, subdiscoidal, sinistral; spire slightly raised; the last whorl with an obscure keel; axis umbilicated, keeled on the edge; operculum? LIMACINA, Cuvier.

Shell thin, vitreous, discoidal, depressed, sinistral; axis umbilicated; whorls smooth; aperture angulated below or canaliculated, sometimes prolonged into a spine-like curved beak; operculum glossy, thin, transparent, of few whorls, with a central muscular scar.

SPIRIALIS, Eydoux and Souleyet.

Genus **LIMACINA**, Cuvier.
Regne Anim., ii. 380. 1817.

1. L. HELICINA, Gmelin. Fig. 253.
(*Clio.*) Syst. Nat., 3149. 1790.

Shell subglobose, subdiscoidal; spire slightly raised; whorls six, last large, with a very obscure keel; axis umbilicated, keeled on the edge.

Diam. 10 mill.

Arctic Seas.

Genus SPIRIALIS, Eydoux and Souleyet.
Rev. Zool., 235. 1840.

1. S. GOULDII, Stimpson. Fig. 254.
Proc. Bost. Soc. Nat. Hist., iv. 8. 1851.
Heterofusus balea, Binney (not Möller), Gould's Invert., edit. ii. 505. 1870.

Shell ovate-globose, vitreous, very thin, pellucid, very light, narrowly and deeply umbilicated; spire conoid; whorls seven, sculptured by minute, distant, unimpressed, revolving lines; last whorl large; aperture about equalling the spire, obtuse in front.

Length 2.5, breadth 1.8 mill.

Massachusetts Bay.

2. S. ALEXANDRI, Verrill. Fig. 255.
Am. Journ. Sci., 3d ser. iii. 281. 1872.
Heterofusus retroversus, Binney (not Fleming), Gould's Invert., edit. ii. 505. 1870.
Spirialis Flemingii? A. Agassiz (not Forbes), Bost. Proc., x. 14. 1865.

Body-whorl very ventricose; spire of four whorls, not forming half the length of the shell.

Nahant, Mass.

Family CLIONIDÆ, Gray.
Syn. Brit. Mus. 1840.

Genus CLIONE, Pallas.
Spicil Zool., x. 28. 1774.

1. C. LIMACINA, Phipps. Fig. 256.
(*Clio.*) Voy. North Pole, 195. 1774.
Clio borealis, Brug. Encyc. Meth. Vers., i. 506. 1792.

Gelatinous, pellucid, pale-blue; mouth and end of the body scarlet when out of water, hyaline; wings somewhat triangular; tail acute.

Portland, Me., northwards.

CLASS ACEPHALA.

Animal without head, always aquatic, contained within a bivalve-shell, one valve of which is applied to the right and the other to the left side of the body.

Synopsis of Families.

A. *Animal provided with siphons.*
a. *Pallial impression sinuated*

Pholadidæ. Shell free; valves equal, gaping at both ends, thin, white, brittle, armed in front with rasp-like imbrications, without hinge-teeth, and strengthened externally by accessory valves; hinge plate reflexed over the beaks, and furnished with a long, curved, tooth-like process beneath each; anterior muscular impression on the hinge-plate; pallial sinus very deep; living perpendicularly in holes in rock or sand.

Gastrochænidæ. Animal symmetrical, elongated; with two long contractile siphons posteriorly, united nearly to their extremities, which are fringed with cirrated orifices. Shell: Valves thin, gaping, edentulous, ligament external, adductor impressions two, pallial line sinuated; contained within a shelly tube, both valves free, or one or both valves cemented to its walls.

Burrowing in wood, stone, sand or mud at low water mark, and lining the burrow with a calcareous tube.

Teredidæ. Animal worm-like; siphons furnished at their extremities with two shelly styles or palletes; shell contained in a shelly tube, globose, its valves trilobate, gaping anteriorly and behind, without hinge-teeth or accessory valves; hinge-plate reflexed over the beaks, and furnished with a long tooth-like process beneath each; living in burrows in wood, which they line with a calcareous tube.

Anatinidæ. Often inequivalve, thin; interior nacreous; surface granular; ligament external, thin; cartilage internal, placed in corresponding pits and furnished with a free ossicle; muscular impressions faint, the anterior elongated; pallial line usually sinuated.

Saxicavidæ. Shell equivalve, solid, gaping at each end; hinge-teeth rudimentary; cartilage external, thick, prominent; pallial impression irregular, sinuated posteriorly; perforating stones or imbedded in sand or mud.

Myidæ. Shell thick, strong, opaque, porcellanous, gaping posteriorly, valves usually unequal, covered with a wrinkled epidermis; hinge simple, toothless, but with a hollow process of the beak in one valve, containing the ligament.

Solenidæ. Shell elongated, gaping at the ends; ligament external; large, lineal, marginal, supported on a prominent pad or fulcrum; hinge-teeth usually 2-3, compressed, the posterior bifid; usually living buried vertically in the sand.

MACTRIDÆ. Shell equivalve, trigonal, close or slightly gaping; cartilage contained in a deep triangular pit of the hinge-plate; hinge with two diverging cardinal teeth, and usually with anterior and posterior laterals; pallial sinus short, rounded.

TELLINIDÆ. Shell free, regular; hinge with two cardinal teeth, at most, in each valve, sometimes lateral teeth; ligament external or internal, on the shorter side of the shell; pallial impression largely and deeply sinuated.

VENERIDÆ. Shell regular, closed, suborbicular or oblong; ligament external; hinge with usually three diverging teeth in each valve; muscular impressions oval, polished; pallial line sinuated.

b. Pallial line simple, without sinus.

CYPRINIDÆ. Shell regular, equivalve, oval or elongated; valves close, solid; epidermis thick and dark; ligament external, conspicuous; cardinal teeth 1-3 in each valve, and usually a posterior lateral tooth; pedal scars close to, or confluent with, the adductors; pallial line simple.

LUCINIDÆ. Shell orbicular, free, closed; hinge-teeth 1 or 2, laterals 1-1 or obsolete; muscular impressions 2, elongated, rugose; ligament inconspicuous or subinternal.

CARDIIDÆ. Shell regular, equivalve, free, cordate, ornamented with radiating ribs; posterior slope sculptured differently from the front and sides; cardinal teeth two, laterals 1-1 in each valve; ligament external, short and prominent; muscular impressions subquadrate.

CHAMIDÆ. Shell inequivalve, thick, attached; beaks subspiral; ligament external; hinge-teeth 2 in one valve, 1 in the other; adductor impressions large, reticulated; pallial line simple.

B. Animal without Siphons; Pallial line simple.

ARCADÆ. Shell free, regular, equivalve, with strong epidermis; hinge with a long row of similar, comb-like teeth; pallial line distinct; muscular impressions subequal.

MYTILIDÆ. Shell free, equivalve, obliquely oval or elongated, closed, umbones anterior, epidermis thick and dark, often filamentose; ligament internal, submarginal, very long; hinge without teeth, or with very minute teeth; anterior muscular impression small and narrow, posterior large, obscure.

AVICULIDÆ. Shell inequivalve, very obliquely resting on the smaller (right) valve, and attached by a byssus; posterior muscular impression large, subcentral, anterior small, within the umbo; pallial line irregularly dotted; hinge line straight, elongated; umbones anterior, eared, the posterior ear wing-like; cartilage contained in one or several grooves; hinge edentulous or obscurely toothed.

OSTRÆIDÆ. Shell inequivalve, slightly inequilateral, free or adherent, resting on one valve; beaks central, straight; ligament internal ; epidermis thin; adductor impression single, behind the centre; pallial line obscure; hinge with or without primary teeth.

Family PHOLADIDÆ.

Synopsis of Genera.

* Anterior hiatus always open. PHOLADINÆ.

Dorsal valves placed anterior and posterior to the beaks, umbonal processes reflected over the beaks, closely applied. PHOLAS.

Dorsal valves lanceolate, placed side by side. Umbonal processes reflected over the beaks, cellular beneath. DACTYLINA.

Destitute of accessory valves. ZIRPHÆA.

** Anterior ventral gap closed in the adult by a callous plate.

JOUANNETINÆ.

With a single accessory valve. MARTESIA.

With two accessory valves, the principal plate over the umbones, with a smaller anterior one adjoining. DIPLOTHYRA.

Genus **PHOLAS**, Linnæus.
Syst. Nat., edit. x. 1758.

There are but four recent species of Pholas known, as now restricted, and they are very easily distinguishable from each other.

* Margins of the valves regularly rounded anteriorly. P. COSTATA.

** Anterior ventral margin emarginate. P. TRUNCATA.

1. P. COSTATA, Linnæus. Figs. 257–259.
Syst. Nat., edit. xii. 1111. 1767.

Shell very large, thin, inflated, with strong crenulate radiating ribs, about half an inch apart on the basal margin, armed with vaulted scales caused by the elevation of growth striæ.

Vertical axis 2, transverse 7 inches.

Whole Coast from New Bedford, Mass., to West Indies.

Subgenus **CYRTOPLEURA**, Tryon.
Monog. Pholadacea, 73. 1862.

Margin of the valves emarginate anteriorly, making a short, wide hiatus.

2. P. TRUNCATA, Say. Figs. 260, 261.
Journ. Philad. Acad., ii. 321. 1822.

Shell subpentangular; anterior obtusely rostrated, wedge-shaped in the middle; posterior margin broadly truncated at the tip; valves transversely wrinkled, crossed by striæ, muricated anteriorly by small erect scales, which form ribs from the beak to basal margin.

Vertical axis 1, transverse 2.5 inches.

Whole Coast from Sable Island to W. Indies; also W. Coast of S. A.

Genus **DACTYLINA**, Gray.
Proc. Zool. Soc., 187. 1847.

In the typical form of this genus the nuclei of the dorsal valves are situated at their outer margins, posterior to the centre; the valves are much emarginate anteriorly, forming a short, wide hiatus. I have characterized as a subgenus, a form of which our American species is the type, as follows:—

Subgenus **GITOCENTRUM**, Tryon.
Monog. Pholadacea, 75. 1862.

Nuclei of the dorsal valves anterior, situated nearer the inner margin; dorsal plates marked by radiating lines; valves not emarginate anteriorly, but regularly rounded; hiatus long and narrow.

1. D. CAMPECHENSIS, Gmelin. Figs. 262, 263.
(*Pholas.*) Syst. Nat., 3216. 1790.
Pholas oblongata, Say, Journ. Philad. Acad., ii. 320. 1822.
Pholas Cundeana, Orb. Moll. Cuba, 215, t. 25, f. 18, 19.
Pholas Chiloensis, King, Zool. Journ., v. 334. 1832.

Shell thin, white transversely, much elongated; basal and hinge margins nearly parallel; anterior and posterior margins rounded; valves transversely and longitudinally striated, the striæ muricated and elevated on the anterior side into costæ, which are more prominently and densely muricated; hinge callus polished, minutely striated transversely and longitudinally, and having about twelve cells.

Transverse axis $4\frac{1}{3}$, vertical $1\frac{1}{4}$ inches.

So. Carolina to West Indies.

Very common on the southern coast, penetrating compact mud or clay. It cannot be distinguished from *D. Chiloensis*, King, a species inhabiting the west coast of South America.

Genus ZIRPHÆA, Leach.

Gray, Ann. and Mag. Nat. Hist., 2d ser. viii. 385. 1851.

1. Z. CRISPATA, Linnæus. Figs. 265, 265, and 266.

(*Pholas.*) Syst. Nat., edit. xii. 1111. 1767.

Pholas semicostata, H. C. Lea, Bost. Journ. Nat. Hist., 285, t. 24, f. 1.

Shell large, thick and strong, oval-oblong, rounded behind, sub-angular or beaked in front; both extremities widely gaping, the valves touching only at the hinge and middle of the basal margin; surface divided into two portions by a broad furrow, running almost vertically from the beaks to the base; the anterior portion coarsely marked with lamellar concentric plates; within smooth, but showing the outer broad vertical furrow; soiled grayish-white, occasionally rust-colored.

Transverse axis 2.5, vertical 1.5 inches.

Northern United States. (*Eur.*)

Subfamily JONANNETINÆ, Tryon.

Proc. A. N. S. Philad. 1862.

Genus MARTESIA, Leach.

Blainville, Dict. Sc. Nat. 1824.

1. M. CUNEIFORMIS, Say. Figs. 267, 268.

(*Pholas.*) Journ. Philad. Acad., ii. 322. 1822.

Wedge-shaped; anterior margin nearly closed, transversely truncated from the hinge; posterior margin with a rounded lip; a deep furrow from the beak to the middle of the basal margin, impressed within; surface with transverse undulating striæ, with elevated minutely crenate lines; hinge callus, forming a cavity before, and without cells; dentiform process filiform, incurved; hinge plate ovate-triangular, with a short projecting angle on the anterior middle, and subacute behind; white.

Transverse axis 20, vertical 11.25 mill.

(Penetrating wood.) *Southern Coast; New Haven, Conn.* (*Perkins.*)

Genus **DIPLOTHYRA**, Tryon.

Proc. Philad. Acad. 1862.

1. D. SMITHII, Tryon.   Fig. 269.

Proc. Philad. Acad. 1862.

Shell short, ovate, divided in the middle by an oblique impressed line, posterior to which the surface is covered with growth lines only, but anteriorly it is finely and sharply transversely sculptured, and obsoletely, radiately ribbed in some specimens; the umbonal plates are generally much distorted, so that no particular form can be traced throughout all the specimens, though the more perfect approach to that depicted in the magnified figure.

Transverse axis 15, vertical 10 mill.

(Burrowing in oyster shells.)   *Tottenville, Staten Island, N. Y.*

Family GASTROCHÆNIDÆ, Gray.

Zool. Proc. London. 1858.

Genus **ROCELLARIA**, Fleuriau de Bellevue.

(*Rupellaria.*)   Journ. de Physique., liv. 1802.

Gastrochæna (partim) Auct.

Shell regular, equivalve; valves ovate or cuneiform, widely gaping anteriorly, very unequilateral; umbones anterior, ligament long and narrow; pallial line lightly impressed, sinuated, uniting the muscular impressions; tube claviform or irregular, often incomplete, perforating shells and limestone, to which its walls are sometimes adherent.

1. R. OVATA, Sowerby.   Fig. 270.

(*Gastrochæna.*)   Zool. Proc. 21. 1834.

Shell ovate, whitish, longitudinally striate, striæ narrow; anterior length one-fifth that of the posterior side.

Length 30, alt. 17.50 mill.

Charleston, S. C.   (W. Stimpson.)

This species inhabits the West Indies, and also occurs on the Pacific side of Central America.

2. R. STIMPSONII, Tryon.   Fig. 271.

Proc. Philad. Acad. 1862.

Shell narrowly elongate, white, anterior extremity very short, acuminate; valves densely concentrically striate; umbones scarcely prominent, nearly terminal; hiatus narrowly elongate, nearly extending the total length of the shell; dorsal and ventral margins nearly parallel.

Length 16, alt. 6 mill.

Beaufort, N. C. (W. Stimpson.)

Family TEREDIDÆ, Carpenter.
Lectures on Mollusca, 100. 1861.

The shelly tube of the *ship-worm* is subcylindrical, divided longitudinally and often concamerated by numerous, incomplete, transverse partitions; the siphonal palettes or stylets assist in compressing and relaxing the siphons to facilitate the flow of water through the long canal.

The Teredines live in most seas, perforating wood in the direction of the grain by means of the mechanical attrition of their valves; these tortuous perforations are lined by calcareous matter forming the tubes. The animal is useful in destroying fragments of wrecks and floating timber, but causes great destruction to dikes, wharves, and to ships when the timbers are not protected from its ravages.

Synopsis of Genera.

Pallets simple. TEREDO.

Pallets compound, the blade penniform, composed of a number of jointed setæ. XYLOTRYA.

Genus TEREDO, Linnæus.
Syst. Nat., edit. x., p. 651. 1759.

1. T. DILATATA, Stimpson. Figs. 272, 273, and 274.
Bost. Proc., iv. 113. 1851.

Diameters nearly equal; wing large, not ascending so high as the beak, but passing off from it by a gentle slope, descending below the anterior triangle, having no defining exterior groove, slightly concave and then reflected outwards on the internal face; pallets very small, battledore-shaped; end of tube concamerated.

Length and alt. 12.5 mill.

Massachusetts to So. Carolina.

This species differs from *T. megotara*, Hanley, which it greatly resembles, in the smaller altitude of the valves, the greater breadth of the auricle, which is also placed much lower, and in its concamerated tubes.

9

2. S. NAVALIS, Linnæus. Figs. 275–280.
     Syst. Nat., edit. x. 651. 1758.

Valves about equal in length and breadth, the posterior auricle, expanded somewhat laterally, its base extending lower than that of the anterior area; anterior area moderate, the basal margin convex, inclining somewhat obliquely downwards to the fang, its junction considerably higher up than that of the posterior auricle ; posterior auricle not ascending, but produced laterally, its dorsal edge mostly somewhat concave, lateral margin nearly straight, a little oblique, rounded at each end; fang acuminating rapidly towards the base ; internally, the apophysis is broad but thin, not thickened at the end, and of the same breadth throughout, and the position of the posterior auricle is defined by a close, projecting rim ; pallets convex on one side and plane on the other, the stalk, which is about as long as the blade, moderately thick and flexous, not continued as a rib beyond the commencement of the blade ; tube not concamerated.

Valves and pallets each 6 mill. in length.

New England and Middle States. (Eur.)

3. T. MEGOTARA, Hanley. Figs. 281–283.
     Brit. Conch., i. 77, t. 1, f. 6, t. 18, f. 1, 2.

Breadth and altitude of valves subequal ; the posterior auricle large, broadly rounded on the margin, raised above the beak and terminating below much further down than the anterior auricle ; the auricle is not defined within by a projecting shelf ; pallets spoon-shaped, with truncate apex, with slender, cylindrical stalks, on the concave side forming a rib to the apex of the blade.

New England. (Eur.)

4. T. NORVAGICA, Spengler. Figs. 287–291.
     Skrivt., Nat. ii., 102, t. 2, f. 4–6, 1792.

Valves solid, higher than wide, the beak elevated beyond the dorsal margins of the auricles ; auricles small, terminating at less than one-half the length of the shell.   Pallets spade-shaped, truncate at the tip with a stem about as long as the blade ; scarcely defined on the centre of the latter.   Tube solid, concamerated.

Height 18, breadth 13 mill.

New England (rare, *Eur.*).

This is the largest and most solid species occurring on our coast ; its great proportionate length, elevated beaks, and the concavely sloping dorsal line of its auricles, as well as the large and peculiar shaped pallets, will readily distinguish it from its allies.

5. T. THOMSONII, Tryon. Figs. 284–286.
     Proc. Philad. Acad., 280, t. 2, f. 3, 4, 5, 1863.

Valves higher than wide; anterior auricle moderate, obliquely subtriangular; posterior auricle small, not very wide, short, somewhat reflected outwards, its dorsal margin does not extend so high as the beak, nor its basal margin so low as that of the anterior area. Within, the posterior auricle is defined by a strong, sharp projection, making a deep sulcus on the exterior surface. Pallets obliquely, or sometimes regularly obovate; style short and directed backwards; margins of the blade convex. From the style an elevated ridge extends around a portion of each side of the blade, and is smooth, while the centre portion, extending to the end, is lunately striate. Tube not concamerated.

Three or four feet below low water mark.

Provincetown and New Bedford, Mass.

Differs from the other species in the very small proportionate size of the posterior auricle, and in its not extending basally as low down as the anterior area.

6. T. CHLOROTICA, Gould. Figs. 292–294.
     Invert. Mass., edit. ii. 33, f. 360, 1870.

Shell minute, subglobose, greenish-white, anterior area very large, posterior area quite small, scarcely defined. Pallets with lyre-shaped blades, the extreme two-thirds covered by a dark encrustation which terminates in two projecting horns. Tube lined by a thin gummy or horny coat, and terminating in a concave calcareous disk with a sort of transverse scar on its outer or convex face.

Diameter 3 mill.

From timbers of ships that have cruised in the Pacific. The tubes penetrate the timber *across the grain of the wood.* On account of its several peculiarities of structure, Dr. Gould proposes for this shell the generic name of *Lyrodus.* Mr. Jeffreys believes this species to be identical with *T. pedicellata,* Quatrefages.

Genus **XYLOTRYA**, Leach.
Menke, Syn. Meth. 1830.

1. X. FIMBRIATA, Jeffreys.   Figs. 295–297.
Ann. and Mag. Nat. Hist., 3d ser. vi. 126.

Shell subtrigonal, diameters about equal, striæ on anterior area about thirty; posterior auricle large, sloping from the beak and descending much below the anterior triangle, inner face fan-shaped, large, overhanging, concave, concentrically striate. Pallets oar-shaped, the blade as long as the handle.

Diam. 6 mill.   Pallets 12 + mill. long.

New Bedford, Mass., Fort Macon, N. C.

The valves resemble those of *T. navalis* so closely that they are scarcely distinguishable from that species. The pallets are, however, entirely different.

Family ANATINIDÆ.

Shell thin, generally inequivalve; interior nacreous; external surface granular; ligament external, thin; cartilage internal, placed in opposite pits and furnished with a free ossicle; muscular impressions faint, the anterior elongated; pallial line usually sinuated.

Animal with mantle-margins united; siphons long, more or less united, fringed; gills single on each side, the outer lamina prolonged dorsally beyond the line of attachment..

Synopsis of Genera.

Oval; inequivalve, left valve deepest; posterior side very short and contracted; beaks fissured, strengthened within by oblique diverging ribs; hinge with a spoon-shaped process in each valve containing the cartilage and ossicle.   PERIPLOMA, Schum.

Subtriangular; inequivalve, fragile; hinge having a narrow ledge within each valve, to which is attached the ligament and an adhering four-sided ossiculum.   LYONSIA, Turton.

Oblong; nearly equivalve, slightly compressed, attenuated, and gaping posteriorly, smooth or minutely scabrous; cartilage processes thick, not prominent, with a crescentic ossicle.   THRACIA, Leach.

Inequivalve, transverse, thin; valves close, attenuated behind; right valve flat, with a diverging ridge and cartilage furrows; left valve convex, with two diverging grooves at the hinge.   PANDORA, Brug.

Genus **PERIPLOMA**, Schum.
Essai Nov. Gen. 115. 1817.
Cochlodesma, Couthouy. Bost. Journ. Nat. Hist., ii. 170. 1839.

1. P. PAPYRACEA, Say. Fig. 298.
(*Anatina.*) Journ. Philad. Acad., ii. 314. 1822.
A. fragilis, Totten. Am. Journ. Sci., xxviii. 347, f. 1.

Shell thin, fragile, rounded-ovate; one valve more convex and at the basal margin projecting a little beyond the other. Beaks not prominent, in the posterior third of the length of the shell; from the beaks to the posterior margin runs an elevated angular ridge; posterior margin narrowed and subtruncated, slightly gaping. Exterior surface minutely wrinkled. Tooth long, narrow, and oblique, with an accessory process at the base. White and pearly.

Height 12, breadth 17 mill.

Whole Coast (rare).

Mr. T. A. Conrad (Am. Journ. Conch., ii. 106) revives Totten's name for the New England shell which he considers to differ from Say's species—the latter being described from the Southern Coast. The outline varies considerably in different specimens, and there does not seem to be sufficient ground for the separation proposed

2. P. LEANA, Conrad. Figs. 436, 437.
(*Anatina.*) Journ. Philad. Acad., vi. 263, t. 11, f. 11.
Cochlodesma Leana, Couthouy. Bost. Journ. Nat. Hist. ii. 170.

Shell very thin and fragile, ovate, subcompressed; the left valve almost flat, rounded at both ends; the right valve convex and subtruncate at the shorter end, slightly gaping at both ends. Beaks small, slightly cleft at one side; from the beaks proceeds a ridge, more or less obvious to the posterior end. Surface wrinkled, with a yellowish shining epidermis extending somewhat beyond the margins; the spoon-shaped process in the hinge nearly horizontal, and resting on an oblique rib directed backwards; no ossiculum.

Vertical axis 22.5, transverse axis 32.5 mill.

Laminarian. *Whole Coast.*

I do not acquiesce in the separation of this species from *Periploma*, as I do not find sufficient distinctive characters for a different genus. The absence of the ossiculum does not appear to be an important character, inasmuch as it is also wanting to some species of *Thracia*, although the typical species possesses it.

Genus LYONSIA, Turton.
Brit. Bivalves, 35. 1822.

Osteodesma, Deshayes, Encyc. Meth., iii. tab. 1830.

The animal has a closed mantle; a tongue-shaped, grooved foot, byssiferous; very short siphons, which are nearly united, fringed; large lips and narrow, triangular palpi.

Distribution, twelve species—all seas.

1. L. HYALINA, Conrad.   Fig. 301.
(*Mya.*) Journ. Philad. Acad., vi. 261, t. 11, f. 12.

Shell thin, fragile, pellucid, transversely elongated; anterior side short, rounded; posterior side produced, compressed, truncated, and reflexed at the end; beaks prominent, inclined forwards; epidermis dirty-white, membranaceous, concentrically wrinkled and corrugated by radiated lines.

Vertical diam. 8.7, transverse 15 mill.

Whole Coast.

2. L. ARENOSA, Möller.   Fig. 302.
(*Pandorina.*) Ind. Moll. Grœnl. 1842.

Shell ovate-quadrate, ventricose, opaque-white; beaks anterior; epidermis finely radiately ridged, with frequently adhering particles of fine sand.

Height 7.5, length 12.5 mill.

New England, northwards.

Distinguished from *L. hyalina* by its smaller size and quadrate form.

Genus THRACIA, Leach.
Blainv., Dict. Sc. Nat., xxxii. 347. 1824.

The mantle of the animal is closed; foot linguiform; siphons rather long, separate, with fringed orifices; gills single, thick, plaited; palpi narrow, pointed.

About twenty species have been described, from northern and temperate zones, and ranging from 4 to 110 fathoms.

1. T. CONRADI, Couthouy.   Figs. 308, 309.
Bost. Journ. Nat. Hist., ii. 153, t. 4, f. 2.

T. declivis, Conrad (not Pennant).   Am. Mar. Conch. 44, t. 9, f. 2.
T. inflata, J. Sowerby.   1845.   (Teste Jeffreys.)

Shell thin, ventricose, rounded in front, narrowed and subtruncate behind. Beaks prominent, with an obtuse carination extending to the angle of the posterior and basal margins; beak of the right valve perforated to receive the point of the left. Right valve more convex and extending somewhat beyond the left, and both slightly gaping. Hinge toothless, but with rounded eminences. Epidermis thin, with concentric undulated striæ. Pallial impression with an acute angular sinus. No ossiculum.

New England, northwards.

2. *T.* MYOPSIS, Beck. Fig. 303.

Möller, Index Moll. Grœnl. 18. 1842.

T. Couthouyi, Stimpson, Bost. Proc., iv. 8. 1851.

Shell small, white, orbicular-ovate, compressed; beaks nearly median, narrowed and rounded in front, more pointed and truncate behind, gaping; surface with rather elevated concentric lines; hinge callus thickened backwards, without any distinct spoon-cavity. Ossiculum very minute.

Height 17.5, length 25 mill.

New England, northwards.

Mr. Jeffreys considers this species a synonym of *T. truncata,* Brown (Brit. Conch. 1827), but I think that Brown's figure represents a very different shell. In thus adopting Brown's name for this species, the following species, *T. truncata,* M. and A., has its name preoccupied, and Mr. Jeffreys proposes to change it to *T. septentrionalis* (Ann. and Mag. Nat. Hist. 238, Oct. 1872). I do not adopt the new name because I am not satisfied that Brown's species is really distinct from *T. distorta,* Montagu.

3. *T. truncata,* Mighels and Adams. Fig. 304.

Bost. Journ. Nat. Hist., iv. 38, t. 4, f. 1. 1842.

Shell small, ovate-triangular, compressed, white, rather solid; beaks at posterior fourth, posterior margin broadly truncate. Beaks small, the right one excavated to receive that of the left valve. Epidermis yellowish, within white. Ligament rather large and prominent. Hinge callosity not spoon-shaped, produced.

Height 12.5, length 19 mill.

New England, northwards.

See remarks under preceding species.

Genus PANDORA, Bruguiere.

Encyc. Meth. t. 250. 1792.

Animal with mantle closed, except a small opening for the narrow, tongue-shaped foot; siphons very short, united nearly throughout, ends diverging, fringed; palpi triangular, narrow; gills plaited, one on each side, with a narrow dorsal border.

Distribution very extensive, burrowing in sand and mud, 4 to 110 fathoms.

1. P. TRILINEATA, Say.   Figs. 305–307.
     Journ. Philad. Acad., ii. 261. 1822.

Shell white, subpellucid, concentrically wrinkled; wedge-shaped, rounded and short before, elongated into a recurved subtruncated beak behind. Hinge-margin concavely curved; surface flattened and bounded on its edges by two elevated lines from the beaks to the rostrated tips. Three or four distinct lines radiate from the beaks. Flat valve with two teeth, of which one is shorter and more robust than the other; the corresponding cavities in the other valve bounded by three tooth-like elevations. Iridescent within.

Height 15, length 30 mill.

Whole Coast.

<h3 style="text-align:center">Family SAXICAVIDÆ.</h3>

Animal symmetrical, oblong. Mantle-lobes united and thickened in front; siphons large, elongated, often invested with a thick, wrinkled epidermis, united nearly to their ends, the orifices fringed; pedal opening small. Foot small, digitiform, inferior, furnished with a byssal groove.

Perforating stones, or living imbedded in sand and mud.

Synopsis of Genera.

Shell oblong, equivalve, valves rugose, gaping, beaks prominent. Hinge when young, with two small teeth in each valve, when adult, edentulous; ligament external, more or less prominent. Muscular impressions strong, wide apart; pallial line interrupted, sinuated posteriorly.   SAXICAVA.

Shell transversely oblong, equivalve, gaping at both ends, surface nearly smooth or transversely furrowed. Hinge with a single conical tooth in each valve, lodged in a cavity of the opposite valve; ligament short, external, prominent, attached to strong ridges. Pallial line interrupted, with a deep posterior sinus.   PANOPÆA.

Shell transversely oblong, equivalve, thick, gaping widely at both ends, valves covered with a thick, black, horny epidermis which extends beyond their edges. Hinge callous, edentulous; ligament large, prominent, external. Muscular impressions wide apart, the hinder elongate; pallial line irregular, strongly marked, the posterior sinus very small.   CYRTODARIA.

Shell elongated, cylindrical, gaping at each end ; epidermis dark, horny, extending beyond the margins ; umbones posterior ; hinge edentulous ; ligament concealed ; pallial line obscure. SOLEMYA.

Genus SAXICAVA, Fleuriau de Bellevue.
Bull. Soc. Philom., No. 62. 1802.

1. S. ARCTICA, Linn. Figs. 310, 311, 312, 313.

(*Mya.*) Syst. Nat., edit. xii. 1113. 1767.

Mytilus rugosus, Pennant, Brit. Zool., iv. 110. t. 63, f. 72. 1777.

Mytilus pholadis, Müll. Zool. Dan. t. 87, f. 1, 2, 3.

Saxicava distorta, Say, Journ. A. N. S., ii. 318. 1822.

Shell irregularly oblong oval, the right valve projecting over the left except at the shorter end, generally gaping ; beaks prominent, from which diverge two ridges or elevated lines, one running near the posterior dorsal margin, the other to the lower angle ; these lines are more or less distinct, or obsolete. Surface coarsely wrinkled ; epidermis thin, dingy yellow.

Length 1 inch, height 15 mill.

Whole Coast. (*Occurs throughout the world.*)

This protean species cannot be described with any accuracy, being modified by the substances into which it bores. It is of universal distribution, and has, under its different aspects, received no less than five generic and fifteen specific names.

In the last edition of Gould's " Invertebrata of Massachusetts" the *S. arctica* and *S. rugosa* are separated as distinct species, but the differences pointed out are not permanent, and are therefore unreliable.

Genus PANOPÆA, Menard.
Ann. du Mus., ix. 131. 1807.

This genus, of about a dozen species, inhabits from low water to 90 fathoms, and from northern seas to Mediterranean Sea and Australia.

1. P. NORVEGICA, Spengler. Figs. 314, 315.

(*Mya.*) Skrivt. Nat. Selsk., iii. 46, t. 2, f. 18.

Glycimerus arctica, Lamarck. Anim. s. Vert., edit. 2, vi. 70.

Shell oblong, trapezoidal, thick, covered with a dark, rough epidermis ; beaks anterior, the anterior and posterior margins oblique and subparallel ; surface raised into two rounded, broad elevations which proceed from the beaks to the basal angles, dividing the surface into three nearly equal portions.

Length 62.5, height 40, diam. 30 mill.

New England, northwards. (*Eur.*)

2. P. BITRUNCATA, Conrad.   Fig. 321.

(*Glycimeris.*)   Proc. Philad. Acad., 216, t. 7, f. 1.   1872.

P. Americana, Conrad.   Coues in Proc. Phil. Acad., 139.   1871.

Shell short, rhomboidal, ventricose, contracted, and obliquely truncated anteriorly; posterior margin oblique, slightly emarginate, cardinal tooth in right valve small, compressed, flattened on the posterior side; pallial sinus widely and obtusely rounded.

Fort Macon, N. C.

A single valve only was received by Mr. Conrad; and I suspect that it is from a submarine fossil deposit, although Mr. Conrad thinks it recent.

Genus CYRTODARIA, Daudin.
Journ. de Phys.   1799.

The animal is larger than the shell, subcylindrical; mantle closed, siphons united, protected by a thick envelope; orifices small; pedal opening small, anterior; foot conical; palpi large, striated inside, the posterior border plain; gills large, extending into the branchial siphon.

There are two species, extensively distributed through the Arctic Seas.

1. C. SILIQUA, Chemnitz.   Figs. 316, 317.

(*Mya.*)   Conch. Cab., xi. 192, t. 198, f. 1934.

Glycimerus incrassata, Lamarck.   Syst. An. s. Vert., 126.

Transversely oblong, compressed, heavy and solid; epidermis thick, shining, obliquely wrinkled; beaks not prominent, eroded; ligament large, prominent on the shorter end.   Interior with a very thick callus in the course of the pallial impression; callus of the hinge broad and prominent.   Shining black; under the epidermis ashen gray.

Length 3.5, height 1.5 inches.

Massachusetts, northwards.   (*Eur.*)

Genus SOLEMYA, Lamarck.
Anim. s. Vert., v. 488.   1818.

The mantle lobes are united behind, with a single, hour-glass shaped, cirrated, siphonal orifice; foot proboscidiform, truncated and fringed at the end; gills forming a single plume on each side, with the laminæ free to the base; palpi long and narrow, nearly free.

The animal is very active, leaping and swimming rapidly.   The leap is performed by contracting the foot at the same time that water is expelled from the posterior opening by closing the valves.

There are four known species ; burrowing in mud, in about two fathoms water.

1. S. VELUM, Say.   Fig. 319.

Journ. Philad. Acad., ii. 317.   1822.

Very thin and fragile, transversely oblong-elliptical ; beaks not elevated, umbones scarcely apparent ; basal and hinge margins parallel, ends rounded. Epidermis glossy, with radiating lines, the edges fringed. Reddish brown, with light yellow radiations.

Length 1, height .5 inch.

New England.

2. S. BOREALIS, Totten. Fig. 318.

Am. Journ. Science, xxvi. 366, f. 1.

Shell fragile, oblong, but larger and more solid than the preceding. Radiations with a larger free space ; the edges of the epidermis not rounded by the slits but preserving a square form, and curved outwards ; the cartilage support not arched or vaulted, but forked, with the hinder part directed obliquely forwards. Dark, blackish-brown.

Length 2.5, height .8 inches.

New England.

Family MYIDÆ.

Animal with the mantle almost entirely closed ; pedal aperture and foot small ; siphons united, partly or wholly retractile ; branchiæ two on each side, elongated.

The shells gape usually at each extremity, and the cartilage is contained in a spoon-shaped cavity at the hinge. Living in the sand or mud, lying on the side.

Synopsis of Genera.

Shell oblong, inequivalve, gaping at the ends, left valve smallest, with a large, flattened, spoon-shaped cartilage process ; pallial sinus large.

MYA.

Shell thick, inequivalve, gibbous, closed, produced posteriorly ; right valve with a prominent tooth in front of the cartilage pit ; left valve smaller, with a projecting cartilage process ; pallial sinus slight ; pedal scars distinct from the adductor impressions.   CORBULA.

Shell globular, attenuated and gaping behind ; right valve a little smaller than the left valve ; umbones strengthened internally by a rib on the posterior side ; cartilage process spatulate, in each valve, with an obsolete tooth in front, and a posterior lateral tooth ; pallial sinus very shallow.

NEÆRA.

Genus MYA, Linnæus.

Syst. Nat., edit. x. 670.   1758.

The animal has a small, straight, linguiform foot; the siphons are combined and covered with a partially retractile. epidermis ; orifices fringed, branchial opening with an inner series of large tentacular filaments; gills not prolonged into the siphon ; palpi elongated, free.

About ten species are known, of universal distribution. They are found in sand or mud, especially estuaries, and ranging, generally, from low water to 25 fathoms.   Our species are edible.

1. M. ARENARIA, Linnæus.   Fig. 322.

Syst. Nat., edit. xii. 1112.   1767.

Mya acuta. Say.   Journ. Philad. Acad., ii. 313.   1822 (*Young*).

Mya mercenaria, Say.   Ibid.

Shell transversely ovate, subequilateral, convex, gaping at both ends, but more so at the posterior end, where the valves curve outwards.   Beaks small ; epidermis rough, wrinkled, yellowish.

Length 3 to 5 inches, height 1.5 to 2 inches.

Inhabits the Whole Coast. (*Eur.*)

Common everywhere, burrowing in sand between high and low water ; its residence is readily detected by a small aperture in the sand, through which it ejects a stream of water upon treading hard on the surface.   On many parts of Long Island the hogs are accustomed to root for this species, and follow the tides with unerring sagacity.

2. M. TRUNCATA, Linnæus.   Figs. 320, 325 (animal).

Syst. Nat., edit. xii. 1112.   1767.

Shell subquadrate, truncated behind, where it gapes widely; basal margin irregularly sinuous; epidermis tough and corrugated ; tooth broader than long, with a slightly thickened lobe on the edge; valves convex, beaks moderately prominent; epidermis yellowish.

Length 2.5 to 3.5 inches, height 1.5 to 2.5 inches.

Northern Coast. (*Eur.*)

Externally, this common species resembles the *Panopæa arctica,* but is readily distinguished by its spoon-shaped tooth.

Genus **CORBULA**, Bruguiere.
Encyc. Meth., t. 230. 1792.

Animal with very short, united siphons; orifices fringed; foot thick and pointed; palpi moderate; gills, two on each side, obscurely striated.

An extensive genus, of universal distribution.

1. C. CONTRACTA, Say. Figs. 326, 327.
Journ. Philad. Acad., ii. 312. 1822.

Shell small, solid, convex; valves subequal, shortest and rounded in front, long and pointed behind. Beaks rather prominent, nearly touching each other at their points; basal margin contracted and concave in the middle. Surface with regular equidistant concentric impressed lines and intervening ridges. A prominent ridge runs from the beaks on each side to the posterior basal margin, including a broad space between them; left valve shutting within the other along the basal margin. Epidermis thin. In one valve the tooth is simple, hooked and turned towards the beak; in the other, it is broader than high, projecting at right angles to the valve, with a deep cavity on the posterior side of the base for the reception of the hooked tooth. Epidermis dull brown. Length 10, height 6.25, diam. 5 mill.

New England to Florida.

Genus **NEÆRA**, Gray.
Griffith's Cuvier. 1834.

Animal with closed mantle, lanceolate foot and short, united siphons, branchial one largest, the anal with a membranous valve, both with a few long, lateral cirri. A genus of quite small shells, universally distributed. About twenty-five species are described.

1. N. PELLUCIDA, Stimpson. Fig. 328.
Invert. Grand Manan., 21, fig. 13. 1853.

Shell small, thin, pale white, subovate, swollen anteriorly, and contracted posteriorly into a short but distinct rostrum. Beaks small, tumid, and placed a little before the middle. Surface nearly smooth about the beaks, with irregular, distant striæ of growth near the margin, which become sharp and well marked on the rostrum. Within smooth and glossy, with minute radiating lines across the disk; teeth very minute. Epidermis white, sometimes pale greenish on the beaks and brownish on the rostrum. Length 5, height 3 +, diam. 3 mill.

New England, northwards.

Family SOLENIDÆ.

Animal with a very large, more or less cylindrical foot; siphons short and united (in the typical species, with long shells) or longer and partly separate in the shorter and more compressed genera; gills narrow, prolonged into the branchial siphons.

Usually living buried vertically in the sand.

Synopsis of Genera.

Shell very long, nearly straight, ends gaping, hinge terminal. One primary tooth in each valve. Pallial impression with a short square sinus.
SOLEN.

Shell very long, gaping and rounded at each end, beaks nearly terminal. Hinge with two teeth in one valve, and three in the other. Pallial line with a small, truncate sinus.
ENSIS.

Shell transversely oblong, curved, rounded and gaping at the ends with a rugose epidermis. Beaks sub-central, hinge teeth 2, 3. Sinus of pallial impression very deep, extending beyond the umbo.
SILIQUARIA.

Shell transversely oblong, epidermis polished, ends rounded and gaping. Hinge anterior to the middle, with three teeth in each valve. Pallial line with a short, rounded sinus.
SILIQUA.

Genus SOLEN, Linnæus.
Syst. Nat., edit. x. 672. 1758.

1. S. VIRIDIS, Say. Figs. 329, 330.
Journ. Philad. Acad., ii. 316. 1822.

Shell transversely oblong, compressed. Hinge margin nearly straight; basal margin curved; posterior end obliquely truncated, a little reflected and rounded near the base; anterior end rounded. Smooth, with very slight concentric growth lines. Epidermis pale green.

Length 2 inches, height .4 inch.

New Jersey Coast, southwards.

Genus ENSIS, Schumacher.
Essai, Nov. Gen. 143. 1817.

Solen (part) Auct.

1. E. AMERICANUS, Gould. Fig. 331.
Invert. Mass., edit. ii. 42. 1870.

Shell elongated, cylindrical, slightly curved, ends truncately rounded. Epidermis glossy, with a long triangular space marked by concentric growth lines, above and below covered with lines parallel with the dorsal and basal margins. Greenish-olive, the central space faded purple.

Length 6, height 1 inch.

Whole Coast.

Distinguished from *E. ensis* of Europe by its greater proportionate width of valves. It is altogether a more robust and larger species.

Found at low-water mark, and considered excellent food.

Genus **SILIQUARIA**, Schum.
Essai, Nov. Gen., 129. 1817.

1. S. GIBBA, Spengler. Fig. 332.
    (*Solen.*) Skrivt. Nat. Selsk., iii. 104. 1794.
Solecurtus caribæus (Lam.) Conrad, Mar. Conch., t. 4, f. 3.

Shell thick and solid, beaks obtuse and little elevated. Surface with a thick, straw-colored, concentrically wrinkled epidermis.

Length 4, height 1.5 inches.

Cape Cod, Mass., southwards.

2. S. DIVISA, Spengler. Fig. 333.
    (*Solen.*) Skrivt. Nat. Selsk., iii. 96. 1794.
Solen fragilis, Pultney, Dorset Cat., 28, t. 4, f. 5. 1799.
Solen centralis, Say, Journ. Philad. Acad., ii. 316. 1822.
Solen bidens, Chemn. Conch. Cab., xi. 203, t. 198, f. 1939. 1795.

Shell small and delicate, transversely oblong-ovate, compressed, arcuated, equilateral. Epidermis yellowish, with a purple band from the beaks to the basal margin.

Length 1.5, height .5.

Southwards from Cape Cod, Mass.

Genus **SILIQUA**, Mühlfeldt.
Entwurf. 44. 1811.
Machæra, Gould, Invert. Mass., p. 32. 1841.

1. S. SQUAMA, Blainville. Fig. 335.
    (*Solecurtus.*) Dict. Sci. Nat., xlix. 419.
Machæra nitida, Gould, Invert. Mass., 33. 1841.

Shell thick, oblong-ovate, beaks small; epidermis very shining, dark yellowish or greenish-yellow, wrinkled.

Length 3, height 1.25 inches.

Northern Coast, northwards.

2. S. COSTATA, Say. Figs. 334, also 299 and 300.
    (*Solen.*) Journ. Philad. Acad., ii. 315. 1822.

Shell thin and fragile, oval-oblong, much compressed, beaks

very minute.   Surface smooth and diaphanous, the epidermis very shining.   Pale violaceous, passing into olive towards the margins; the violet disposed in faint broadening rays.   Within white, faintly iridescent, the transverse rib white.

Length 2, height .8 inches.

New England to New Jersey.

This species is more southern in distribution than *S. squama;* it is smaller, more fragile, and readily distinguished by its purple rays.

Family MACTRIDÆ.

Animal with the mantle more or less open in front; siphonal tubes united, orifices fringed; foot compressed; gills not prolonged into the branchial siphon.

Synopsis of Genera.

Shell thick, nearly equilateral; anterior hinge-tooth **Λ** shaped, with sometimes a small laminar tooth close to it ; lateral teeth doubled in the right valve.   MACTRA.

Shell thin, cordate, ventricose, slightly produced and rather gaping behind; hinder slope keeled, narrow.   RAETA.

Genus **MACTRA**, Linnæus.
Syst. Nat., edit. xii. 1125.   1767.

The mactras inhabit sandy coasts, where they bury just beneath the surface ; the foot can be stretched out considerably and moved about like a finger, it is also used for leaping.   The animal is eaten by star-fishes, whelks, pigs, and men.

1. M. SOLIDISSIMA, Chemnitz, 336, 337.
   Conch. Cab., x. 350, t. 170, f. 1656.

Mactra similis, Say, Journ. Philad. Acad., ii. 309.   1822.
Mactra Raveneli, Conrad, Am. Mar. Conch., 65.   1831.

Shell large, solid, subovate, or subtriangular; cardinal fosset very large, cordate; lateral teeth transversely striated ; muscular impressions very large ; epidermis dirty-brown.

Whole Coast.

American authors have generally considered the *M. similis* distinct from the above.   It is said to be smaller in size, more triangular in outline, and to replace the *solidissima* on the Southern coast.   Having collected these shells in large numbers at Atlantic City, N. J., a locality where northern and southern forms intermingle, I have satisfied myself that *similis* is not distinct; but I figure the latter species (Fig. 337) so that collectors can form their own conclusions. *M. Raveneli* is a more transverse variety, but does not seem to possess any distinctive characters.

This is the largest of our bivalve shells, attaining, in northern specimens, a length of seven inches, and great solidity.

2. M. OVALIS, Gould. Figs. 340, 341.
  Invert. Mass., edit. i. 53. 1841.
Mactra polynyma, Stimpson, ~~Shells of N.~~ Eng. 20. ~1851. *S.J. Checklist eos. 3.*

Shell large, thick, obovate, coarse, nearly equipartite, covered with a tough, dusky-brown epidermis; V tooth strong, lateral teeth not striated; sinus of pallial impression deep.

New England to N. Carolina.

This species is readily distinguished from *solidissima* by its shorter, more ventricose form, and the smooth lateral teeth. It is somewhat smaller in size, yet becomes quite ponderous with age, attaining a length of over four inches. Fig. 341 represents the young of this species, described by Stimpson as *M. polynyma.*

3. M. LATERALIS, Say. Figs. 338, 339.
  Journ. Philad. Acad. ii., 309. 1822.

Shell triangular, very convex, of a smooth appearance, but with minute concentric wrinkles; lateral margins flattened cordate, with a rectilinear, sometimes concave profile; one margin rounded at the extremity, the other longer and less obtuse; umbo nearly central, prominent.

Length 22, height 20, diam. 15 mill.

Whole Coast.

This is more triangular and ventricose than *M. ovalis;* the flexuous and obtusely ridged posterior margin also distinguishes it.

4. M. FRAGILIS, Chemnitz. Fig. 342.
  Conch. Cab., vi. 236, t. 24, f. 235.
Mactra oblonga, Say. Journ. Philad. Acad., ii. 310. 1822.

Shell oblong-oval, very slightly wrinkled, excepting upon the margin; umbo hardly prominent, from which a carinated angle passes to the posterior extremity; epidermis pale brownish-yellow, tinged with ferruginous; within white, highly polished.

North Carolina to West Indies.

10

Spurious Species.

Mactra nucleus, Conrad. This species, described from two odd valves stated to have been collected at Long Branch, N. J., is now ascertained to be a Manilla species.

Genus RAETA, Gray.

1. R. CANALICULATA, Say.   Fig. 343.
   Journ. Philad. Acad., ii. 311.   1822.

Shell transversely oval orbicular, very thin and fragile, inflated; valves with equal concentric grooves; posterior margin short, subreniform, compressed; a marginal longitudinal irregular sub-impressed line, between which and the edges the grooves become mere wrinkles; posterior slope nearly straight; gape considerable; anterior margin regularly curved; within grooved. White.

Length 2.5, height 2 inches.

New Jersey, southwards.

2. R. LINEATA, Say.   Fig. 344.
   Journ. Philad. Acad., ii. 310.   1822.

Shell transversely sub-oval, thin; posterior gap patulous; anterior linear, and commencing beyond the hinge slope; valves unequally obsoletely wrinkled; anterior margin smooth with a carinated line. White.

Length 2.75, height 2 inches.

New Jersey, southwards.

Family TELLINIDÆ.

Animal with the mantle widely open in front, its margins fringed; foot tongue-shaped, compressed; siphons separate, generally very long and slender; palpi large, triangular; gills united posteriorly, unequal, the outer pair sometimes directed dorsally.

The shell is free, compressed, usually closed and equivalve; cardinal teeth two at most, laterals 1–1, sometimes obsolete; muscular impressions rounded, polished; pallial sinus very large; ligament on the shortest side of the shell.

The Tellens are found in all seas, chiefly in the littoral and laminarian zones; they frequent sandy bottoms, or sandy mud, burying beneath the surface.

Synopsis of Genera.

* *Ligament external.*

Shell ovate, oblong or rounded posteriorly; beaked or angular, with a flexuosity on the hind slope. One or two primary teeth in each valve; lateral teeth present or obsolete.		TELLINA.

Shell orbicular, convex, surface of valves divaricately striated; posterior flexure obsolete. A small anterior and a large bifid cardinal tooth in the right valve, and a single cardinal tooth in the left valve; lateral teeth two in each valve.		STRIGILLA.

Shell oval, convex or sub-ventricose. Cardinal teeth small; no lateral teeth.		MACOMA.

Shell triangular, very inequivalve, the right valve concave; surface of valves plicate; beaks acute, laterally incurved; lateral slopes strongly produced and dentate at their edges. Hinge with two primary teeth in one valve and one in the other; lateral teeth two in each valve.

TELLIDORA.

Shell oblong or irregular; two cardinal teeth in the right valve, the posterior one thin and directed obliquely backwards; two cardinal teeth in the left valve, the posterior one stout, bilobed, the anterior one smaller. No distinct lateral teeth.		GASTRANELLA.

Shell more or less wedge-shaped, equivalve; the hinder side much shorter than the anterior. Two cardinal teeth in one valve and one bifid tooth in the other; one or two lateral teeth in each valve.		DONAX.

* *Cartilage internal, situated in a cartilage pit.*

Shell thin, transversely elongated, slightly gaping at the sides; surface smooth, covered with a thin, deciduous epidermis. Cartilage pit oblique; primary teeth small or wanting; lateral teeth distinct; ligament short, partly external.		ABRA.

Shell transversely oval or orbicular, slightly gaping at the sides. Hinge with one or two primary teeth in each valve, with a long narrow pit between them for the internal cartilage; ligament external, thin.

SEMELE.

Shell equivalve, inequilateral, anteriorly rounded, posteriorly subtruncate and slightly gaping. Hinge with a small, anterior primary tooth in each valve; cartilage internal, in a spoon-shaped cavity projecting into the cavity of the valves; one strong lateral tooth on each side of the hinge in one valve, no lateral teeth in the other.		CUMINGIA.

Shell ovate cuneate, truncated behind. Hinge with a simple, compressed primary tooth and a rudimentary process in the place of the second tooth; lateral teeth sub-equal, compressed, strongly cross-grooved. CERONIA.

Shell equivalve, inequilateral, oblong, closed; surface nearly smooth, or concentrically striated. Two diverging teeth in each valve, one of them in the right valve, elevated and conspicuous; cartilage in a pit in each valve; lateral teeth none. Muscular impressions strong; pallial sinus large and broad.   ERVILIA.

Genus Tellina, Linnæus.

Syst. Nat., edit. x. 674. 1758.

1. T. ALTERNATA, Say.   Fig. 345.

Journ. Acad. Nat. Sci. Philad., ii. 275. 1822.

Shell compressed, oblong, narrow and angulated behind; numerous impressed concentric lines, alternately obsolete, on the posterior margin. Within, a callous line passes from behind the hinge to the inner margin of the posterior cicatrix. Posterior hinge tooth emarginate; anterior lamellar tooth near the cardinal, so as to appear like a primary tooth; that of the right valve wanting; posterior lamellar tooth at the extremity of the ligament. Posterior hinge-slope declining in a concave line to an obliquely truncated tip. White, tinged with yellow within.

Length 55, height 31 mill.

North Carolina to West Indies.

2. T. POLITA, Say.   Figs. 346, 347, 348.

Journ. Philad. Acad., ii. 276. 1822.

Shell transversely subtriangular, with minute concentric wrinkles; anterior margin rather shortest; hinge slope declining in a very slightly arcuated line to a subacute termination; basal margin nearly straight from before the middle to the posterior end; a lateral tooth behind the primary one. White immaculate.

Length 15, height 10 mill.

North Carolina, southwards.

3. T. TENERA, Say.   Fig. 349.

Journ. Philad. Acad., ii. 303. 1822.

Shell very thin and fragile, pellucid, compressed, transversely oblong, suboval; covered by delicate concentric lines of growth. Beaks placed slightly posteriorly; marginal folds distinct; basal margin slightly arcuated. Posterior cardinal tooth in the left valve largest; the other often indistinct; the chief tooth in each valve grooved; lateral tooth on the longest side distinct; the others very indistinct. White, iridescent, occasionally rosaceous.

Length 14, height 9 mill.

Nova Scotia to S. Carolina.

Found abundantly on sandy beaches, and probably lives not far from low-water mark.

4. T. TENTA, Say.   Figs. 350, 351.

    American Conchology, Pl. 65, f. 3.

Shell small, oval, white; shortest behind, narrowed, much deflected and widely gaping; lines of growth very fine. Two cardinal teeth in the right and one in the left valve; a posterior lateral tooth in the right valve and a corresponding groove in the left.

    Length 15, height 10 mill.

Massachusetts to So. Carolina.

This species is generally larger than *T. tenera*, and has not its polished surface. It differs from others principally in being less triangular, more strongly deflected, and in its widely gaping posterior portion.

5. T. MODESTA, Verrill.   Figs. 352, 353.

    (*Angulus.*)   Am. Journ. Sci., 285, t. 6, f. 2, 2a.   1872.

Shell smooth, shining, more or less iridescent, with very fine concentric striæ. Form similar to *T. tenera*, but more oblong and with the anterior dorsal margin nearly straight or even slightly concave; beaks at about the posterior third and scarcely prominent; the posterior end slopes rapidly and is subtruncate at the end; the ventral margin is but slightly convex in the middle, and subparallel with the dorsal margin. Teeth and hinge margin stronger than in *tenera*. Color pink, light straw-color or white; often banded concentrically with these colors.

New England.

6. T. IRIS, Say.   Fig. 354.

    Journ. Philad. Acad., ii. 302.   1822.

Shell very thin and fragile, pellucid, compressed, transversely oblong, suboval; minute concentric wrinkles, crossed by oblique striæ which do not attain the margin; margin narrowed and subacute; basal edge straight, opposite the beaks. Color, white, iridescent, with a rosaceous disk and one or two anterior rays.

    Length 12.5, height 7.5 mill.

North Carolina, southwards.

7. T. BREVIFRONS, Say.   Figs. 355-357.

    Am. Conchology, t. 64, f. 1.

Shell oval, thin, and fragile, not very convex, white, tinged with pale dull fulvous; with transverse slender striæ and obsolete radiating ones; deeper colored within; cardinal teeth two in the left valve and one in the right valve; lateral teeth none.

South Carolina.

A doubtful species, of which only one specimen was received by Mr. Say.

8. T. DECORA, Say.　Figs. 358, 359.
　　Journ. Philad. Acad., v. 219.　1826.

Transversely subovate, not much compressed, with numerous minute concentric wrinkles and regular equidistant lines crossing them; no oblique lines on the posterior margin. Anterior lateral tooth of the left valve prominent, the others obsolete; apex a little before the middle. Rosaceous, or white with rosaceous radiations.
　Length 22, height 19 mill.

　　　　　　　　　　　　　　　　　Southern Coast.

9. T. LATERALIS, Say. = Cayennense Lam,
　　Journ. Philad. Acad., v. 218.　1826.

Shell transversely subovate; beaks nearly central. Anterior margin regularly rounded; posterior margin rostrated, the beak turning to the left and slightly gaping; ligament slope straight; basal margin regularly arcuated. Valves with small concentric wrinkles and slight waves; within, these are slightly impressed. Lateral teeth none; cardinal teeth two in one valve, and one, with another scarcely elevated filiform tooth in the other. Whitish, often tinged with rusty; within white.
　Length 52, height 37 mill.

　　　　　　　　　　　　　　　　　Southern Coast.

I am not acquainted with this species.

10. T. LUSORIA, Say.　Fig. 360.
　　(*Psammobia.*)　Journ. Philad. Acad., ii. 304.　1822.

Shell oblong, suboval, with minute wrinkles; posterior side narrowed, and inclining to the right at the end; an obtuse convex line on the left valve.　Bluish-white.
　Length 1, height .6 inch.

　　　　　　　　　　　　　　　　　New Jersey to Florida.

　　　　　　　　　　Doubtful Species.

11. T. VERSICOLOR, Cozzens.　Fig. 361.
　　DeKay, Moll. N. Y. 209, t. 26, f. 272.　1843.

Shell transverse, compressed, inequilateral, slightly gaping at its subacute extremity, smooth, posterior end subangular, with an indistinct fold; anterior extremity dilated and rounded. Cardinal teeth two in the right valve; the posterior more robust, simple, in the left valve rudimentary or inconspicuous. White, opalescent, with a purple or bluish iridescence often in the form of rays.

Length 17, height 10 mill.

New York.

Genus STRIGILLA, Turton.
Brit. Bivalves, 117. 1822.

This genus is readily recognized by its obliquely-sculptured valves and its orbicular shape.

1. S. CARNARIA, Linnæus. Figs. 362, 363.
   (*Tellina.*) Syst. Nat., edit. xii. 1119. 1767.

Suborbicular, convex, strong, glossy, pink or rose color; inside rosy; lateral teeth distinct and nearly equidistant.

Southern Coast to W. Indies.

2. S. FLEXUOSA, Say. Figs. 364, 365.
   (*Tellina.*) Journ. Philad. Acad., ii. 303. 1822.
Strigilla mirabilis, Phil. Wiegm. Archiv., 260. 1841.

Shell suborbicular, white, smaller and more convex than *S. carnaria*, the flexuous lines more numerously angled.

Length 12, height 11 mill.

North Carolina to W. Indies.

Doubtful Species.

3. S. MERA, Say. Fig. 366.
   (*Tellina.*) Am. Conch., vii. t. 64, f. 2.

Shell ovate-orbicular, concentrically striated, hinge bidentate, with a lateral tooth in one valve.

South Carolina.

The above description and figure refer to a species which has not been identified; its analogies of form would seem to indicate its position in this genus. Mr. Say remarks that "in a particular light it has a slight appearance of longitudinal lines." It is quite possible that it is a worn specimen of *Strigilla* in which the zigzag oblique lines are obsolete.

Genus MACOMA, Leach.
Journ. de Phys., lxxxviii. 465. 1819.

1. M. BALTHICA, Linnæus. Fig. 367.
   (*Tellina.*) Syst. Nat., edit. xii. 1120. 1767.
Psammobia fusca, Say, Journ. Philad. Acad., v. 220. 1827.

Shell thin and fragile, ovate orbicular, beaks small, almost central. Surface with concentric wrinkles; rounded before and somewhat pointed behind ; white or pink, covered with a dusky epidermis.   Teeth, two in each valve, slender and slightly diverging, the largest grooved.

Length 25, height 20 mill.

Whole Coast, Arctic Seas, N. Europe.

2. M. CALCAREA, Chemnitz.   Fig. 368.

(*Tellina.*)   Conch. Cab., vi. 140, f. 136.   1782.

Tellina sabulosa, Spengler, Skrivt., Nat. iv., Pt. 2.   1798.
Tellina proxima, Gray, Zool. Beechey's Voy., 154, t. 44, f. 4.   1839.
Tellina sordida, Couthouy, Bost. Journ. Nat. Hist., ii. 59, t. 3, f. 11.

Shell thin and fragile, inequilateral, subtriangular, slightly gaping.   Epidermis thin and brittle; beneath which the surface is marked with numerous incremental lines.   Beaks very small and behind them the margin slopes away in nearly a straight line.   Teeth two in each valve, the largest bifid.   White, covered by a thin, brownish epidermis.

Length 22, height 15 mill.

Arctic Seas to New York.

Wider and more triangular and depressed than the preceding species.

3. M. SUBROSEA, Conrad.   Fig. 369.

Am. Journ. Conch., vi. 71, t. 1, f. 3.   1871.

Subtriangular, equilateral, convex ; substance very thin ; beaks slightly prominent, direct; posterior side cuneiform ; ventral margin rounded, disk minutely striated concentrically, white or rosaceous and glossy, with a thin, pale ochreous epidermis ; cardinal tooth in the left valve compressed, with a minute linear sulcus.

Raritan Bay, N. J. ; Delaware Bay.

Doubtful Species.

4. M. TENUIS, Da Costa.   Fig. 370.

(*Tellina.*)   Conrad, in Say's American Conchology, vii. t. 64, f. 3.

Shell oval-triangular, irregularly striate concentrically ; each valve with two teeth and one of them with lateral teeth.

Sullivan's Island, S. Car.

The above meagre description and figure represent a shell sent to Mr. Say by Prof. Ravenel of Charleston, S. C., which Mr. Conrad has identified as *M. tenuis*.   I am not aware that it has since been found on our coast.

Genus **TELLIDORA**, Mörch.

1. T. LUNULATA, Holmes.   Figs. 371–3.

Post Plioc. Fossils S. Car., 47, t. 9, f. 7.   1860.

Shell subtriangular, inequivalve, inequilateral; surface of valves concentrically plicated; beaks prominent, slightly curved laterally; lateral slopes strongly produced, dentated at their edges; two primary teeth in the right valve, one in the other; lateral teeth two in each valve.

No. Carolina. southwards.   (Living.)

Genus **GASTRANELLA**, Verrill.

1. G. TUMIDA, Verrill.   Figs. 374, 375.

Am. Journ. Science, iii. 286, t. 6, f. 3.   1872.

Shell small, variable in form, swollen above, more or less elongated, oval or oblong, with rounded ends, compressed posteriorly. Beaks rounded, somewhat prominent, incurved, but not approximate, and directed somewhat forward; anterior dorsal margin deeply concave in front of the beaks, but without a distinct lunule, at the anterior end regularly rounded or a little prolonged, compressed; ventral margin slightly convex, or nearly straight and subparallel with the dorsal margin, or incurved, in the different specimens; posterior end broadly rounded in some, decidedly prolonged in others; dorsal posterior margin usually nearly straight for at least half its length, sometimes a little convex and gradually sloping throughout. Surface with fine somewhat irregular, concentric striæ, slightly iridescent. White, with the umbones purple.

Long Island Sound, near New Haven, Conn.

This species appears to be a "nestler" and quite variable in form. About 20 specimens were obtained of different sizes; one of the largest, which may not be mature, is .18 of an inch long, .09 high, and about the same in thickness. In 4–6 fathoms, shelly and gravelly bottom, among hydroids and sponges.

Genus **DONAX**, Linnæus.

Syst. Nat., edit. x. 682.   1758.

1. D. FOSSOR, Say.   Figs. 376, 377.

Journ. Philad. Acad., ii. 306.   1822.

Donax angustatus, Sowerby.   Thes. Conch. Monag. Donax.

Shell subtriangular, anterior margin short and rounded. Posterior hinge-slope straight; the base very slightly prominent beyond a regular curve at the middle; basal margin crenate within; pale livid, with or without obscure rays.

Length 12.5 mill.

New Jersey.

This very distinct species appears to be extremely local in distribution; it does not occur north of Long Island Sound, nor has it been detected south of Delaware Bay.

2. D. VARIABILIS, Say. Figs. 378, 379.

Journ. Philad. Acad., ii. 305, 1822.

Shell triangular; anterior margin obliquely truncated, cordate, suture a little convex; posterior hinge margin nearly straight; base a little prominent beyond a regular curve near the middle. Valves striated longitudinally with scarcely visible parallel impressed lines; basal edge crenate. White or bluish with rays of bluish-purple.

Length 22, height 12 mill.

Beaufort, N. C., southwards.

Genus ABRA, Leach.

Lam. Anim. s. Vert. 1818.

1. A. ÆQUALIS, Say. Figs. 380, 381.

(*Amphidesma.*) Journ. Philad. Acad., ii. 307. 1822.

Shell orbicular, slightly oblique, polished, white, with very minute and numerous concentric wrinkles near the margin, which are obsolete on the disk and umbo; lateral teeth none; primary teeth two in the left valve and one in the other; interior ligament cavity subfusiform, as long as the exterior ligament.

Length 10 mill.

New Haven, Conn., North Carolina, southwards.

Genus SEMELE, Schumacher.

Essai Nov. Gen., 165. 1817.

1. S. ORBICULATA, Say. Fig. 382.

(*Amphidesma.*) Journ. Philad. Acad., ii. 317. 1822.

Shell orbicular, somewhat compressed; beaks nearly central and a little prominent; valves slightly wrinkled concentrically; hinge with two lamellar teeth. White.

Length and height 27 mill.

North Carolina, southwards.

2. S. RADIATA, Say. Fig. 383.

(*Amphidesma.*) Journ. Philad. Acad., v. 220. 1826.

Transversely oval orbicular, a little compressed. Apex nearly central, a little prominent, posterior slope slightly concave. Primary teeth two in each valve; lateral teeth very distinct. White with rosaceous radiating bands, sometimes obsolete; within tinged with yellow, and showing the bands.

Length 27, height 23 mill.

Georgia, southwards.

3. S. NEXILIS, Gould.
Proc. Bost. Soc. Nat. Hist., viii. 280. 1862.

Shell oblong-oval, white, with a blush towards the umbones; reticulated by concentric laminar striæ and remote radiating ribs, their intersections punctate, and muriculate towards the ends; umbones a little behind the middle.

Length 25, height 15 mill.

Coast of Georgia.

The sculpture is similar to that in the young of *Lucina tigerina*, but much more delicate.

The above is from Gould's description; I have not seen the species.

4. S. ORNATA, Gould.
Proc. Bost. Soc. Nat. Hist., vii. 280. 1862.

Shell small, elliptical, reddish, rosaceous near the margin in radiating and concentric lines; concentrically costate-striate and radiately striate; within flavous somewhat radiated with red.

Coast of Georgia.

This species I have not seen. It was dredged by the U. S. Coast Survey.

Genus CUMINGIA, Sowerby.
Proc. Zool. Soc., 3, 4. 1833.

The species of *Cumingia* are usually found in sponges, sand, and fissures of rocks; the valves, in consequence, often assume an irregular aspect.

1. C. TELLINOIDES, Conrad. Fig. 384.
(*Mactra.*) Journ. Philad. Acad., vi. 258, t. 9, f. 2, 3. 1830.

Shell ovate-triangular, thin, white, nearly equilateral; pointed and deflected behind, broadly rounded before; surface covered by numerous sharp, elevated growth lines.

Length 15, height 11 mill.

New Bedford, Mass., to Georgia.

Genus CERONIA, Gray.

1. C. ARCTATA, Conrad.   Figs. 385, 386.

   (*Mactra.*)   Journ. Philad. Acad., vi. 257, t. 11, f. 1.   1830.

Mactra deaurata, Conrad, Am. Mar. Conch. 59, t. 14, f. 1.

Mactra subtriangulata, Wood, Index Suppl. t. 1, f. 10.

Shell subtriangular, wedge-shaped, thick and strong, smooth and covered by a thin yellowish epidermis. Hinge with a V-shaped primary tooth and a long straight lateral tooth on either side, partially double in the left valve and their articulating surfaces striated.

Length 37, height 25 mill.

New York to Labrador.

2. C. DEAURATA, Turton.   Fig. 387.

   (*Mactra.*)   Conch. Dithyra Brit., 71, t. 5, f. 8.   1822.

Mactra denticulata, Gray, in Wood Suppl., t. 1, f. 9.

Mesodesma Jauresii, Joannis, Mag. de Zool., t. 54.   1834.

Shell ovate, triangular, thick, very rough with coarse concentric ridges ; lateral teeth very strong, curved, faintly striated.

Length 43, height 26 mill.

New Foundland, Gulf of St. Lawrence.

This species is larger, more ovate in form, flatter, rougher than *C. arctata*, and differs internally in its curved lateral teeth and their obsolete striation.

Genus ERVILIA, Turton.
Brit. Bivalves, 56.   1822.

1. E. CONCENTRICA, Gould.

   Proc. Bost. Soc. Nat. Hist., viii. 280.   1862.

Shell minute, oblong-ovate, pellucid, shining, crowded with concentric raised striæ; umbones a little posterior, anterior more acute than the posterior extremity.

Length 6+, height 4, diam. 3 mill.

Dredged off the Coast of North Carolina.

This little shell, which seems to be abundant along the whole Southern coast, is quite different from anything before described. —GOULD.

Family VENERIDÆ.

Animal free, locomotive, rarely byssiferous or burrowing; mantle with a rather large anterior opening; siphons unequal, more or less united; foot linguiform, compressed, sometimes grooved; palpi moderate, triangular, pointed; branchiæ large, subquadrate, united posteriorly.

Synopsis of Genera.

Shell ventricose, subglobose, triangularly heart-shaped; valves with the margins finely crenulated. Hinge with three erect compressed diverging teeth in each valve, the anterior in the left and posterior in the right valve strong and somewhat bifid, the others simple and lamellar; lozenge prominent, obliquely sulcately crenate within. Pallial line remote from the ventral margin, ending behind in a short, narrow, triangular sinus.

MERCENARIA.

Shell roundly triangular, subequilateral; surface of valves smooth, the margins crenulated. Three primary teeth in the left valve, the middle one conical and slightly arched, and two in the right valve, diverging, with a wide pit between them. Muscular impressions ventral; pallial line marginal, with a very long, narrow, deep sinus ascending perpendicularly.

GEMMA.

Shell ovately triangular, more or less thickened or subcordiform; margins of the valves finely crenulated. Hinge narrow, solid, tridentate in the right valve, bidentate in the left; teeth divaricate, unequal, the anterior tooth the longest. Pallial line with the sinus nearly obsolete, or very short and triangular.

CHIONE.

Shell transverse, ovate, inequilateral; margins of the valves entire, often obtuse. Hinge tridentate in the left valve, the anterior lateral tooth united to the lunular tooth. Pallial line with a wide, deep, semiovate sinus.

CALLISTA.

Shell orbicular, compressed, concentrically striated, deeply lunulate under the beaks. Hinge with three teeth in each valve, the lunular tooth elongate and compressed; ligament external, partially concealed under the lozenge. Sinus of pallial impression deep, oblique, triangular, with the apex acuminate.

DOSINIA.

Shell transversely ovate, inequilateral, margins entire. Hinge tridentate, teeth sometimes diverging, sometimes approximate, subparallel, often bifid or canaliculate at the apex. Pallial impression deeply sinuated posteriorly; sinus semioval, somewhat horizontal.

TAPES.

Shell oblong or ovate, white, covered with a hard, thin epidermis, ventricose, anterior side short, posterior gaping. Hinge composed of two primary teeth in each valve, one of which is often obsolete or rudimentary; lateral teeth none; ligament short, external. Pallial line with a deep rounded sinus.

PETRICOLA.

Genus **MERCENARIA**, Schumacher.
Essai Nov. Gen., 135. 1817.

1. M. MERCENARIA, Linnæus.   Figs. 388, 389, 390.
(*Venus*.)   Syst. Nat., edit. xii. 1131. 1767.
Mercenaria violacea, Schumacher, Essai, Nov. 135, t. 10, f. 3. 1817.

VARIETY.

Venus notata, Say, Journ. Philad. Acad., ii. 271. 1822.

Shell solid, obliquely ovate, very inequilateral; lunule heart-shaped; surface bluish-white, with numerous concentric, laminated ridges, becoming obsolete on the middle; inner margin stained with violet.

The variety (perfect young or half-grown shells) has the surface covered with fawn-colored zigzag markings, and the interior is wholly white (Fig. 390).

Usual length 3 inches, height 2½ inches.   It sometimes attains much greater proportions.

Massachusetts to North Carolina.

This species is the common *Round Clam*, so much prized as an article of food.   Its aboriginal name of *Quahog* has now fallen into disuse.   It abounds in all our bays, a few inches below the surface, from low-water mark to six fathoms.   If taken from its bed and placed on its side, it can, in the course of a single tide, bury itself six inches.   From the colored internal margin of the shell the *wampum* or colored beads, constituting the specie currency of the Indians, was formerly manufactured.

2. M. MORTONI, Conrad.   Figs. 391, 392.
(*Venus*.)   Journ. Philad. Acad., vii. 251. 1837.
Mercenaria fulgurans, Tryon, Am. Journ Conch., i. 1865.
? *Venus præparca*, Say, Journ. Philad. Acad., ii. 271. 1822.

Shell very large, cordate, inflated, thick and ponderous, with prominent recurved concentric laminæ, more elevated at the sides; ligament margin arcuate.   Umbones prominent; lunule large, cordate, defined by a deep groove; posterior extremity slightly emarginate; cavity of the cartilage profound.   Teeth large, prominent, grooved; muscular impressions very large; inner margin regularly crenulated.

Length 5 to 6 inches.

North Carolina, southwards.

Mercenaria fulgurans, Tryon, (Fig. 391) appears to be a younger state of this species, in which the surface is covered with zigzag fulvous or purplish lines.

Venus præparca, Say, has not been positively identified; it may be the young either of this or the preceding species.

Genus **GEMMA**, Deshayes.

Tottenia, Perkins. Bost. Proc., 148. 1869.

1. G. GEMMA, Totten. Fig. 393.

(*Venus.*) Silliman's Journal, xxvi. 367, f. 2, a.-d. 1834.

Gemma Totteni, Stimson, Check Lists. 1860.

Tottenia gemma, Perkins, Bost. Proc., 148. 1869.

Shell minute, nearly orbicular, equilateral, beaks slightly elevated; concentrically furrowed; violet and white; margin crenulate.

Length 3.5 mill.

New England to North Carolina.

This species is viviparous, producing three dozen young at a time.

2. G. MANHATTENSIS, Prime. Fig. 394.

Ann. New York Lyceum, vii. 482. 1852.

Shell small, triangular, solid, shining; beaks nearly central, elevated; surface grooved with remote concentric furrows, inner margin crenulated.

Length and height 3 mill.

New York.

It is rather smaller, more triangular, and more deeply and regularly grooved than *G. gemma*, and destitute of purple within and without.

Genus **CHIONE**, Muhlfeldt.

Gray, Zool. Proc., 183. 1847.

The animal has short, broad, unequal siphons, united at their bases, the branchial with two rows of cirri, the anal ciliated. Mantle-margins plicato-dentate.

1. C. CINGENDA, Dillwyn. Fig. 395.

(*Venus.*) Desc. Cat. Shells. 1815.

Venus cancellata, Lamarck, Anim. s. Vert. 1818.

Venus elevata, Say, Journ. Philad. Acad., ii. 272. 1822.

Shell subcordate, longitudinally sulcated, sulci equal, numerous, dense, on the anterior submargin sparse, crossed by concentric, elevated, remote, lamellar bands, white, with blotches of red or purple, or bluish-gray.

Length 22.5, height 20 mill.

North Carolina, southwards.

2. C. TRAPEZOIDALIS, Kürtz.

(*Venus*.) Cat. Shells, N. and S. Car. 1860.

Shell covered with convex radiating ribs, set with brown spots and scales of growth. A thin brown pile on good specimens.

Length 12.5 mill.

North and South Carolina.

Fossil Species.

C. ALVEATA (*Venus*), Conrad.

C. INÆQUALIS (*Venus*), Say.

These species are included in Stimpson's Catalogue of Shells of the Atlantic Coast, but I am confident they have not been found except in a fossilized condition.

Genus CALLISTA, Poli.

Test. Sicil., i. 30. 1791.

The mantle margins are plicate, with filaments above the base of the respiratory siphon; siphons united to their ends, crowned with simple cirrhi.

1. C. GIGANTEA, Chemnitz. Fig. 396.

Conch. Cab., f. 1661.

Shell large, ovate, smooth, slightly angulated on the anterior side; posterior depression oblong-ovate, a little impressed on its sides and keeled in the middle. Teeth compressed. Color pale livid with numerous lilac longitudinal broad rays, generally interrupted.

Length 6, height 3.25 inches.

North Carolina, southwards.

2. C. MACULATA, Linnæus. Fig. 397.

(*Venus*.) Syst. Nat., edit. xii. 432. 1767.

Shell oval, rather compressed, posteriorly; obliquely somewhat produced; fawn white, blotched or waved with violet brown, enveloped with a shining horny epidermis.

Georgia to West Indies.

3. C. SAYANA, Conrad. Fig. 398.

(*Cytherea*.) Am. Journ. Science, xxiii. 345. 1833.

C. convexa, Say, of authors.

Shell moderately solid, ventricose, subcordate; beaks elevated, directed forwards. Anterior lunule heart-shaped, distinctly marked by a simple line. Epidermis dingy white.

Length 43, height 35 mill.

New England and Middle States.

The true *C. convexa* of Say is a different species, only occurring fossil.

Genus DOSINIA, Scopoli.

The siphons are united ; mantle-margin plicate ; foot subquadrangular, without a byssal groove.

1. D. DISCUS, Reeve. Fig. 399.

Monog. Conch. Icon., vi. sp. 9.

Artemis concentrica, Conrad (not Gmelin), Mar. Conch., t. 12.

Shell compressed, with fine, regular, impressed concentric striæ ; beaks considerably curved, pointed ; lunule cordate, slightly impressed ; epidermis yellowish-white ; hinge with a large oblong fosset under the beaks ; muscular impressions very large.

Length and height 3.5 inches.

Virginia to West Indies.

Genus TAPES, Mühlfeldt.
Entwurf. 51. 1811.

Siphons united as far as the middle, diverging at their ends ; branchial siphon crowned with arborescent tentacles ; mantle-margin simple ; foot lanceolate, byssiferous.

1. T. FLUCTUOSA, Gould. Figs. 400, 401.

(*Venus.*) Invert. Mass., 1st edit. 87. 1842.

Venus Astartoides, Beck. Middend. Beitr. Mal. Ross., iii. 56. 1849.

Shell transversely ovate, thin, lenticular, white, with a yellowish epidermis ; surface with recurved concentric waves vanishing at the sides ; areola none, or indistinct in old specimens.

Length 20, height 15 mill.

Newfoundland, northward.

Genus PETRICOLA, Lamarck.
Syst. Anim. s. Vert. 121. 1801.

Siphons elongated, distinct, their orifices ciliated ; the mantle-lobes are united except a small anterior opening ; foot compressed, lanceolate, with a byssiferous fissure a little behind the middle of the lower edge.

The Petricolas excavate limestone or coral rocks, and also bury in mud.

11

1. P. Pholadiformis, Lamarck.   Figs. 402, 403.
   Anim. sans Vert., v. 565.   1818.

Petricola fornicata, Say, Journ. Philad. Acad., ii. 319.   1822.
Petricola dactylus, Say (not Sowb.), Am. Conch, t. 60, f. 2.   1834.

Shell elongated, anterior side short, with strong ribs crossed by waved striæ; posterior side with radiating lines and gaping; teeth three in one valve, and two in the other.   White.

Length 37, height 17 mill.

Whole Coast.

Family CYPRINIDÆ.

Animal with the mantle-lobes united posteriorly by a curtain, pierced with two siphonal orifices; foot thick, tongue-shaped; gills two on each side, large, unequal, united behind, forming a complete partition; palpi moderate, lanceolate.

Synopsis of Genera.

Shell large, oval, strong, with usually an oblique line or angle on the posterior side of each valve; epidermis thick and dark; ligament prominent, umbones oblique, no lunule; cardinal teeth 2.2, laterals 0-1, 1-0, muscular impressions oval, polished.   Cyprina.

Shell suborbicular, compressed, thick, smooth or concentrically furrowed; lunule impressed; ligament external; epidermis dark; hinge-teeth 2.2, the anterior tooth of the right valve large and thick; anterior pedal scar distinct.   Astarte.

Shell minute, triangular, furrowed; hinge like *Astarte*, with lateral teeth.
   Gouldia.

Shell rounded or oblong, radiately ribbed; margin toothed; hinge-teeth 1.2, and an elongated posterior tooth; anterior pedal scar close to adductor.   Cardita.

Genus CYPRINA, Lamarck.
Extr. d'un Cours.   1812.

1. C. Islandica, Linnæus.   Figs. 404, 405, 406.
   (*Venus.*)   Syst. Nat., edit. xii. 1131.   1767.

Shell large, thick, and ponderous, ventricose; beaks prominent, incurved, contiguous.   Ligament stout and prominent; basal margin simple, rounded.   Cardinal teeth stout and diverging, lateral inconspicuous.   Epidermis coarse and wrinkled, blackish, becoming olivaceous or brownish towards the margin.

Length 3.3, height 2.8, diam. 1.4 inches.

Massachusetts, northward (Eur.).

Genus **ASTARTE,** Sowerby.
Min. Conch., t. 137. 1816.

Animal with mantle open ; margins plain or slightly fringed ; siphonal orifices simple ; foot moderate, tongue-shaped ; lips large, palpi lanceolate; gills nearly equal, united behind, and attached to the siphonal band.

The genus is Arctic in distribution ; a few species found in warm climates are scarcely typical.

1. A. BOREALIS, Chemnitz. Fig. 407.
Conch. Cab., vii. t. 39, f. 412. 1784.
Astarte semisulcata, Leach, Ann. Phil., xiv. f. 204. Gould, Invert. Mass.. edit. ii. 121. 1870.
Astarte lactea, Gould, Invert. Mass., edit. i. 80. 1841.

Shell orbicular elliptical, compressed, with remote, obtuse, rounded ridges; beaks nearly central; epidermis yellowish-brown, an obsolete lateral tooth in each valve ; margin plain.

Length 27.5, height 25 mill.

New England to Arctic Ocean.

2. A. CASTANEA, Say. Figs. 408, 409, 410.
Journ. Philad. Acad., ii. 273. 1822.

Shell thick and heavy, subtrigonal, with prominent and nearly central beaks, much more elevated than usual in the genus. Surface with minute wrinkles and larger concentric waves, and faint traces of radiating lines. Anterior area very deeply excavated, short, broad, and smooth ; posterior slope almost straight, with a long narrow lanceolate depression. Margin internally crenulated. Epidermis chestnut-brown.

Length and height 1 inch.

New England to New Jersey.

The foot of the animal is a bright vermilion color. The high beaks curved forwards, giving the shell a somewhat kidney-form appearance, will distinguish this species from all others. It is the only species occurring as far south as southern coast of New Jersey, where it is not uncommon.

3. A. COMPRESSA, Montagu. Fig. 411.
(*Venus.*) Test. Brit. Suppl. 43, t. 26, f. 1. 1803.
Astarte striata, Gray. Append. Beechey's Voy., t. 44, f. 9.
Astarte Banksii, Leach, Append. Ross' Voy. 1819.

Shell small, ovate-triangular, elevated, beaks prominent, acute ;

anterior margin concave, with a deep, broad lunule; surface with rather close concentric ribs and striæ, thirty to forty in number, sometimes obsolete towards the base; inner margin simple.

New England, northwards.

4. A. DEPRESSA, Brown.　Figs. 412, 415.

(*Crassina.*)　Brit. Conch. 96, t. 38, f. 2.　1827.

Astarte Warhami, Hancock.

Astarte crebricostata, Forbes, Ann. Nat. Hist., xix. 98, t. 9, f. 4.　1837.

Astarte lens, Stimpson, Verrill, Am. Journ. Science, 287.　Mar. 1872.

Astarte elliptica, Brown, Brit. Conch. 96, t. 38, f. 3.　1844.

Shell large, moderately convex or depressed, beaks rather obtuse, ovate-triangular; anterior slope slightly concave, the posterior end obtusely rounded or truncated; with thirty to forty squared concentric ribs, more or less obsolete towards the posterior end or base.　Margin finely crenate within.

Length 27, height 25 mill.

Maine, northwards.　(Europe.)

Dr. Gould writes of this species, " A series of the young may be selected which would satisfy any one as to specific value; while intermediate specimens would so connect it with *sulcata, elliptica, Banksii,* and *compressa,* as to be a complete puzzle.　A large compressed form, to which I notice that Dr. Stimpson has attached a label in his collection with the name *A. lens,* I think would fall under this species, though it merges almost as well into *A. sulcata.*"　Mr. Verrill distinguishes the American species from *depressa* (*crebricostata*), under the name of *lens;* I am inclined to adopt a more conservative view, in this genus, where the specific characters appear to be so greatly modified in different individuals.

Figures 413–415 represent *A. elliptica,* Brown, which can scarcely be designated as a variety.

5. A. QUADRANS, Gould.　Fig. 416, 417.

Invert. Mass., edit. i. 81, f. 48.　1841.

Astarte Portlandica, Mighels, Proc. Bost. Soc. Nat. Hist., 129.　1843.

Shell obliquely oval, anterior portion the longest; surface smooth, very slightly wrinkled by growth-lines; epidermis light yellowish-olive; hinge margin narrow, with a small lateral tooth in the left valve, and a corresponding groove in the right valve; inner margin plain.

Length 12, height 10 mill.

Massachusetts, northward.

6. A. LUTEA, Perkins.   Fig. 418.

> Proc. Bost. Soc. Nat. Hist., xiii. 151, figure.   1869.

Shell gibbous, thick, subtrigonal, length and breadth nearly equal; beaks prominent, incurved, not meeting; surface with twenty or more concentric ridges; margin crenulated within.

Length 23, breadth 21 mill.

Connecticut.

Animal with light yellow mantle, edged with bright orange ; foot bright orange, striped longitudinally with yellow. Ovaries full of bright orange ova in April.

A somewhat doubtful species ; its form too close to that of *borealis.*

7. A. SULCATA, Da Costa.   Figs. 419, 420.

> (*Venus.*) Brit. Conch., 192. 1778.

Crassina Danmoniensis, Lamarck, Anim. s. Vert., edit. Deshayes, vj. 360.
Astarte undata, Gould, Invert. Mass., 80. 1841.
Astarte latisulca, Hanley, Desc. Cat. 87, t. 14, f. 35. 1843.

Shell ovate-triangular, thick, somewhat compressed ; anterior side somewhat shortest ; beaks in contact, obtusely pointed; surface with from ten to twenty concentric furrows and ridges, the former wider than the latter. Epidermis dark brown. Hinge margin crenulated.

Length 31, height 25 mill.

New England, northwards.

Doubtful Species.

8. A. LUNULATA, Conrad, Foss. Tert. Form., 44, t. 21, f. 8.

> *A. bilunulata*, Conr. Adams. Genera, ii. 484.

This species is inserted in Stimpson's Check-List of Atlantic Coast Shells, but I think it exceedingly improbable that it has been correctly identified with any living species.

Genus GOULDIA, C. B. Adams.

1. G. MACTRACEA, Linsley.   Fig. 421.

> (*Astarte.*) Gould, Am. Journ. Sci., 233. 1849.

Shell small, solid, trapezoidal or quadrant shaped, triangular above, rounded below; surface undulated by about fourteen concentric waves or ribs, with very minute radiating striæ. Color yellowish-green.

Length and height 6 mill.

Massachusetts to South Carolina.

2. G. FASTIGIATA. Gould.
Bost. Proc., viii. 280. 1862.

Shell small, obliquely triangular, rather solid, yellowish, concentrically sharply ribbed; apex acute, anterior margin concave, dorsal margin arcuate; ventral margin nearly straight, anterior angle distinct, posterior angle rounded.

Length and height 8 mill.

Frying Pan Shoals, N. Car.

I have not seen this species, nor has it been figured.

Genus **CARDITA**, Bruguiere.
Encyc. Meth. i. 401. 1789.

Animal with the mantle-lobes free, except between the siphonal orifices; branchial margin with conspicuous cirri; foot rounded and grooved, spinning a byssus, labial palpi short, triangular, plaited, gills rounded in front, tapering behind, and united together, the outer pair narrowest.

Recent systematists have separated a number of genera from Cardita, and generally, with sufficiently good distinctive characters; I have indicated these groups in the specific descriptions.

1. C. BOREALIS, Conrad. Fig. 422.
Am. Mar. Conch., 39, t. 8, f. 1. 1831.
Cardita vestita, Deshayes, Zool. Proc., t. 17, f. 10. 1852.

Shell suborbicular, thick, with about eighteen rounded ribs, and narrow interstices, concentrically striated; epidermis brownish-black; margins crenulated within.

Length and height 1 inch, diam. .7 inch.

New York, northwards.

This is the type of Conrad's genus *Cyclocardia*, which also includes the following species.

2. C. NOVANGLIÆ, Morse. Fig. 423.
(*Cyclocardia.*) First An. Rep. Peabody Acad., 76, f. 1869.

Shell oblong ovate, thin, beaks nearly central, not prominent; with about seventeen ribs and concentric striæ; margin crenate within.

Length 21, height 16 mill.

New England, northwards.

This species is more transverse and thinner than *C. borealis,* the beaks are not so elevated or projecting, and the hinge-plate is much narrower.

3. C. TRIDENTATA, Say. Figs. 424, 425.

(*Venericardia.*) Journ. Philad. Acad., v. 216, Am. Conch. t. 40.

Shell suborbicular, subequilateral, thick and ponderous, with about eighteen convex ribs, crossed by concentric elevated lines; within the margin is deeply crenate; hinge with two diverging teeth. separated by a large cavity in one valve, and in the other a single, large, triangular, recurved tooth, closing into the cavity. Length 6.5, height 6 mill.

South Carolina.

This is a somewhat doubtful species, and appears to have been described from a single specimen, which was possibly young and the hinge not perfect. Can it be the young of *C. Novangliæ*? Conrad has made a genus "*Pleuromeris*" for his *C. tridentata*, a fossil shell which is not specifically identical with Say's species, although it belongs apparently to the same group.

Spurious and Doubtful Species.

C. INCRASSATA, Sowb., Conrad, Mar. Conch., is an East Indian species.

C. (CARDITAMERA) FLORIDANA, Conrad, Fossil Shells, 12. 1837.

Inhabits Key West and Tampa Bay, Florida, but I think it has not been detected on the Atlantic coast, although it is included in Stimpson's Catalogue.

VENERICARDIA CRIBRARIA, Say, cover of Am. Conch., pt. 5.

A single specimen from the coast of New Jersey. Say writes "can this be a variety of the *borealis* of Conrad? Having but a single specimen, I cannot determine this question."

Family LUCINIDÆ.

Animal with mantle-lobes open below, and having one or two siphonal orifices behind; foot elongated, cylindrical, or ligulate, protruded at the base of the shell; gills one (or two) on each side. large and thick, oval; mouth and palpi usually minute.

The Lucinidæ are distributed chiefly in the tropical and temperate seas, upon sandy and muddy bottoms, from the sea-shore to the greatest habitable depths.

Synopsis of Genera.

Shell orbicular, white; umbones depressed; lunule distinct; margins smooth or minutely crenulated; ligament oblique, semi-internal; hinge teeth 2.2, laterals 1-1 and 2-2, or obsolete; muscular impressions rugose, anterior elongated within the pallial line, posterior oblong ; umbonal area with an oblique furrow.   LUCINA.

Shell globular, posterior side furrowed or angulated, umbones much recurved ; lunule short or indistinct ; ligament to a certain extent external, placed in a groove on the hinge-line, and outside the hinge-plate ; teeth altogether wanting.   CRYPTODON.

Shell suborbicular, smooth ; ligament double, rather long, submarginal ; hinge-teeth 2.2, of which the anterior in the left valve and posterior in the right, are bifid, lateral teeth none ; muscular impressions polished, rounded.   MYSIA.

Shell small, thin, suborbicular, closed ; beaks small ; margins smooth ; ligament internal, interrupting the margin, or on the thickened margins ; cardinal teeth 1 or 2, laterals 1-1 in each valve.   KELLIA.

Shell oblong, inequilateral, anterior side very short ; ligament concealed between the valves ; hinge-teeth 2-2.   TURTONIA.

Shell small, thin, oblong, anterior side longest ; hinge line notched ; ligament internal, between two laminar diverging teeth.   MONTACUTA.

Shell equivalve, orbicular, subequilateral, compressed, gaping slightly at the sides ; surface of valves shagreened or smooth ; margins plain ; beaks acute.   Hinge composed of a pair of teeth-like laminæ on each side of a triangular, central excision in one valve ; a primary apical tooth in front of a subtriangular excision of the hinge-margin, and flanked on each side by a sublateral lamina in the other.   Pallial impressions simple.
LEPTON.

Genus LUCINA, Bruguiere.
Encyc. Meth., t. 284. 1792.

Animal with the mantle freely open below ; siphonal orifices simple ; mouth minute, lips thin ; gills single on each side, very large and thick ; foot long, cylindrical, pointed, slightly heeled at the base.

1. L. FILOSA, Stimpson. Fig. 426.
Shells of New England, 17. 1851.
Lucina radula, Gould (not Montagu), Invert. Mass., edit. i. 69. 1841.
Lucina contracta, De Kay (not Say), Nat. Hist. New York, 214, t. 27, f. 275. 1843.

Shell orbicular, depressed ; surface with numerous remote concentric laminated striæ ; lunule depressed lanceolate ; lateral teeth obsolete.   White or light brown.

Length and height 1.5 inches.

New England.

2. L. PUSILLA, Gould.
    Proc. Bost. Soc. Nat. Hist., viii. 282. 1862.

Shell minute, reniform, yellowish, slightly concentrically striate; umbones a little posterior, elevated; anterior dorsal margin excavated, extremity retusely rounded; posterior extremity broadly rounded, subtruncated; within radiatingly striated, the striæ evanescent towards the umbones.
    Length 3, height 2.5 mill.

> Coast of North Carolina. (Coast Survey.)

3. L. DENTATA, Wood. Fig. 427.
    Gen. Conch., 195, t. 46, f. 7. 1817.

Lucina strigilla, Stimpson, Shells N. E., 17. 1851.
Lucina Americana, C. B. Adams, Contrib. Conch., 243. 1852.
Lucina divaricata, Lamarck (not Linn.), Anim. s. Vert., v. 541. 1818.[1]

Shell thin, orbicular, equilateral; beaks small, prominent, inclined forwards; basal margin regularly rounded and crenate. Surface with well-marked concentric-lines, crossed by deep, oblique, narrow furrows, flexed at nearly right angles at the anterior third of the surface. White.
    Length 25, height 22 mill.

> Entire Coast. (Distribution universal.)

4. L. TIGERINA, Linnæus. Fig. 428.
    Syst. Nat., edit. xii. 1133. 1767.

Shell oblong-ovate, longer than high, flatly convex, radiately many grooved, very closely decussated throughout with concentric ridges; white within and without.

> Southern Coast. (W. Ind.)

5. L. EDENTULA, Linnæus. Fig. 429.
    Mus. Ulric, 74.

Lucina chrysostoma, Phil. Zeit., Malak., ii. 181. 1845.

Shell orbicular, rather thin, ventricose, superficially excavated from the umbones on each side, concentrically finely and closely striated, teeth obsolete; semitransparent white, orange within.

> Southern Coast. (W. Ind.)

---

[1] For full synonymy and notes on this species, see TRYON, Proc. Philad. Acad., 85. 1872.

Genus **CRYPTODON**, Turton.

Brit. Bivalves, 121. 1822.

1. C. Gouldii, Philippi.   Fig. 430.

(*Lucina.*)   Zeit., Malak., 74.   1845.

Lucina flexuosa, Gould (not Montagu), Invert. Mass., edit. i. 71, f. 52.

Shell minute, white, ovate-globose, nearly equilateral; beaks prominent, inclined forwards, and having a rounded impression in front of them; surface smooth, white; glossy within, with minute radiating lines.

Length 7.5, height 8 mill.

New England.

2. C. obesus, Verrill.   Fig. 431.

Am. Journ. Science, 287, t. 7, f. 2.   1872.

Shell white, irregularly and rather coarsely concentrically striated, much swollen in the middle; transverse diameter nearly equal to the length; height considerably exceeding the length; beaks prolonged and turned strongly to the anterior side; lunular area rather large and sunken, somewhat flat, in some cases separated by a slight ridge into an inner and an outer portion; anterior border with a prominent rounded angle; ventral margin prolonged and rounded in the middle; posterior side with two strongly developed flexures, separated by deep grooves: interior of shell with radiating grooves, most conspicuous toward the ventral edge.

Length 15, height 18 mill.

British America.

More nearly related to *C. flexuosus* of Europe than to *C. Gouldii.* The latter is thinner and more delicate, more rounded, relatively much longer and much smaller.

Genus **MYSIA**, Leach.

Menke, Syn., edit. ii. 112.   1830.

1. M. punctata, Say.

(*? Amphidesma.*)   Journ. Philad. Acad., ii. 308.   1822.

Orbicular, with numerous minute concentric wrinkles and very numerous minute punctures; within, a small rim or projecting line runs near the edge from the hinge to the basal margin. White.

Length and height 7.5 mill.

Southern Coast.

Genus **KELLIA**, Turton.
Brit. Bivalves, 57. 1822.

The animal has a very short posterior siphon: anal tube undivided, entire below.

1. K. PLANULATA, Stimpson. Fig. 432.
Shells of New Eng., 17. 1851.
Kellia rubra, Gould (not Montagu), Invert. Mass., edit. i. 60. 1841.

Shell minute, rather thick, suboval; beaks prominent, in contact, with a well-defined lunule in front of them; anterior, white, with a thin purplish epidermis.
Length 4, height 3 mill.

New England.

More compressed and longer than the next species.

2. K. SUBORBICULARIS, Montagu. Figs. 433, 434, 435.
(*Mya.*) Test. Brit., 39, 564, t. 26, f. 6. 1803.

Shell quadrangular or rounded, swelled, thin and fragile; white with a very thin somewhat iridescent epidermis; beaks nearly median, small, pointed, inclining inwards rather than forwards; no lunule.
Length and height 8 mill.

New England. (*N. Eur.*)

Genus **TURTONIA**, Hanley.
Brit. Moll., ii. 81. 1849.

Anal siphon slender and produced. Foot large, heeled.

1. T. NITIDA, Verrill. Figs. 438, 439.
Am. Journ. Sci., iii. 286, t. 7, f. 4, 4*a*. 1872.
T. minuta, Gould (not Fab.), Invert. Mass., edit. ii. 85, f. 395. 1871.

Shell minute, ovate, rather convex, fragile, semitransparent, beaks at about the anterior third, elevated, inclined forwards; smooth, straw-colored, blending into dark-purple at the beaks and posterior slope; anterior margin broadly rounded, posterior margin more acutely rounded.
Length 2, height 1.7 mill.

Found in crevices of shells and rocks, and among the roots of sea-weeds.

Northern Coast.

Genus **MONTACUTA**, Turton.
Conch. Dict., 102.   1819.

1. M. ELEVATA, Stimpson.   Fig. 440.
   Shells New Eng., 16.   1851.

Montacula bidentata, Gould (non Auct.), Invert. Mass., edit. i. 59.   1841.

Shell ovate, triangular, beaks tumid, elevated, nearly central, disk flattened below the middle; tooth on the shorter side oblique and excavated for the reception of the ligament.   White, under a very thin straw-colored epidermis.

Length 5, height 4 mill.

New England.

2. M. GOULDI, Thomson.   Fig. 441. =*Lep/ . : aba gella Com*
   Am. Journ. Conch., iii. 33, t. 1, f. 15.   1867.

Shell minute, diaphanous, rhomboidal, inequilateral, not compressed.   Beaks rather prominent, not in contact, with an excavated areola in front.   Basal margin nearly straight, ends obtusely rounded, forming a rhomboidal outline; lines of growth regular, with an opaque white thickened band surrounding the margin; hinge with the cartilage occupying a pit between two rather strong teeth.

New Bedford, Mass.

Genus **LEPTON**, Turton.
Brit. Bivalves, 62.   1819.

The mantle is much produced beyond the margin of the shell, and is furnished with slender, tentacular filaments.   The foot is folded as in *Arca,* but when the animal is in motion it forms an expanded disk.

1. L. LEPIDUM, Say.   Fig. 461.
   (*Amphidesma.*) Journ. Philad. Acad., v. 221.   1826.

Shell very much compressed, subtriangular, remarkably thin, pellucid, equilateral, somewhat iridescent, with numerous concentric wrinkles, and equally numerous, very minute, regular, longitudinal striæ, curving towards the anterior and posterior edges; cardinal teeth obsolete, laterals prominent.

Length 6, height 6.5 mill.

South Carolina.

2. L. LONGIPES, Stimpson.
   Bost. Proc., v. 111.   1855.

Shell subtriangular, somewhat rounded, slightly compressed,

smooth and polished, broad anteriorly and sloping at the posterior dorsal margin; beaks elevated; and lower margin nearly straight.

Animal white, nearly transparent ; mantle extending beyond the margin of the shell, open in front, with undulated but unfringed margins; foot large and powerful and may be expanded to double the length of the shell, with a posterior heel. The animal suspends itself by its foot, or can use it for creeping like a gasteropod.

Charleston, S. C.

Unidentified Species.

3. L. FABAGELLA, Conrad. Figs. 442–444.
   Am. Mar. Conch., 53, t. xi. f. 3.

Shell suboval, convex, with minute crowded concentric lines ; beaks central, rather prominent; epidermis yellowish, very thin, wrinkled ; teeth similar in each valve; the posterior tooth longest, and angulated under the beak.

Rhode Island.

A single specimen obtained.

Family CARDIIDÆ.

Palpi slender, acuminate. Mantle freely open in front; siphons distinct but very short, and nearly sessile, their bases and sides furnished with tentacular filaments ; gills two on each side, thick, united together behind the body. Foot very long and geniculate.

Synopsis of Genera.

Shell globose, gibbose, nearly equilateral, more or less gaping posteriorly, . the margins often serrated ; valves with elevated radiating ribs.
CARDIUM.

Shell longitudinally oval, inequilateral ; surface of valves plain (not ribbed); hinder gap small. LÆVICARDIUM.

Shell subcordate, compressed, rather thin, subequilateral, valves with obsolete, radiating ridges, slightly gaping ; beaks rather prominent. Hinge with the cardinal teeth wanting. SERRIPES, Beck.

Genus **CARDIUM**, Linnæus.
Syst. Nat., edit. x. 1758.

1. C. ISOCARDIA, Linnæus. Fig. 445.
   Syst. Nat., edit. x. 679. 1758.

Shell obliquely heart-shaped, gibbous ; radiately ribbed, ribs about thirty-four in number, squamiferous, scales vaulted, rather elevated, slightly flattened on the posterior side, anterior scales more or less obtusely thickened. Pale straw-color, stained with purple-brown, interior bright purple-scarlet, especially towards the umbones.

Southern Coast. (*West Indies.*)

2. C. MURICATUM, Linnæus. Figs. 446, 447.
  Syst. Nat., edit. x. 680. 1758.

Shell ovate, heart-shaped, with thirty-six ribs, of which twelve have their spines directed in an opposite direction to the others; marginal serratures largest on the anterior edge. Grayish or yellowish-white, edged with orange-yellow or scarlet on the anterior side, and sometimes stained with red.

Length 37, height 40 mill.

North Carolina, southwards.

3. C. ELEGANTULUM, Beck. Fig. 448.
  Mörch. Prodr. Faun. Grœnl., 20. 1857.

Shell small, oval, beaks a little anterior; with twenty-six to twenty-eight ribs, separated by deep, wide grooves, and crossed by imbricated bars.

Length 6, height 5 mill.

Greenland.

4. C. MAGNUM, Born. Fig. 449.
  Test. Mus. Cæs. Vind., 46, t. 3, f. 5. 1780.

Shell very large, obliquely cordate, ventricose, posterior side somewhat angularly depressed; radiately ribbed, ribs about thirty-five in number, flattened, rather close-set, anterior ribs crenulated; yellowish-brown, painted with transverse rows of purple-brown spots, the depressed posterior area entirely purple-brown.

North Carolina, southwards.

5. C. PINNULATUM, Conrad. Fig. 450.
  Journ. Philad. Acad., vi. 260, t. 11, f. 8. 1836.

Shell small, thin, and fragile, obliquely orbicular; ribs about twenty-six, flattened, but becoming convex towards the base, crossed by a series of equidistant flattened scales; beaks slightly elevated, often decorticated, inclining inwards. Dingy white or yellowish.

Length 12.5, height 11 mill.

New England, New York.

6. C. ISLANDICUM, Linnæus. Figs. 451, 452, (*C. pubescens*).
Syst. Nat., edit. xii. 1124. 1767.
Cardium ciliatum, O. Fabricius, Faun. Grœnl., 410. 1780.
Cardium pubescens, Couthouy, Bost. Journ. Nat. Hist., ii. 60, t. 3, f. 6.
(Young.)

Shell large and rather thin, rounded, inflated, nearly equilateral. Beaks prominent, incurved, contiguous; anterior dorsal area feebly impressed, subcordate; surface with from thirty-six to thirty-eight sharp ribs, which are covered with a stiff fringe-like epidermis in the young shells. Epidermis dull yellowish-brown, straw-colored within.

Length 2.3, height 2.5 inches.

Cape Cod, Mass., northwards.

Cardium Hayesii and *C. Dawsonii* of Stimpson are probably only Arctic varieties of this species.

Genus LÆVICARDIUM, Swainson.
Malacol., 373. 1840.

1. L. SERRATUM, Linnæus. Fig. 453.
Syst. Nat., edit. x. 680. 1758.
C. lævigatum, Gmel., Syst. Nat., 3251. 1790.
C. citrinum, Wood, Gen. Conch., t. 54, f. 3. 1817.

Shell ovate, rather gibbous towards the umbones, smooth, shining, anteriorly rather obsoletely striated; whitish-yellow, posterior side bright citron-yellow, sometimes stained with pink towards the margin, yellowish within.

North Carolina to West Indies.

2. L. MORTONI, Conrad. Figs. 454-457.
Journ. Philad. Acad., vi. 259, t. 11, f. 5-7. 1831.

Shell small, thin, inflated, globular, slightly oblique; surface smooth, posterior side somewhat obliquely extended; margin entire or obsoletely serrated; beaks large, tumid, subcentral, contiguous. Color very pale yellowish, covered with a very thin darker epidermis, in young specimens with blotches or zigzag lines of dark fawn color; within yellow, with generally a dark purple blotch along the posterior margin.

Length (adults) 1 inch, height 22, breadth 17 mill.

The animal is white, has short, conical siphons, each marked with a circle of brown spots, and fringed with numerous cirri which extend far beyond the shell.

Whole Coast.

3. L. PICTUM, Ravenel.

Proc. Philad. Acad., 44. 1861.

Shell ovate, triangular, very oblique, somewhat compressed, smooth, polished, with a few obsolete ribs at each end, and obsoletely waved by the lines of growth ; beaks small, prominent, nearly touching, very much in advance of the centre, anterior end short, regularly curved, posterior end produced, somewhat angular. Color reddish-brown in zigzag spots and blotches upon a white ground, internally polished, reddish-brown, clouded, with some patches of yellow and a little white ; margin crenulated.

Length 18, height 20 mill.

Charleston, S. C.

I have not seen this species ; it is, perhaps, a highly-colored *C. Mortoni.*

Genus SERRIPES, Beck.

Verzeich. d. Deutsch. Naturf. in Kiel, 217.

Aphrodite, Lea, Am. Philos. Trans. v. 1834.

1. S. GRŒNLANDICUS, Chemnitz. Fig. 458.

(*Cardium.*) Conch. Cab., vi. t. 19, f. 198. 1782.

Aphrodite columba, Lea, Trans. Am. Philos. Soc., v. t. 18, f. 54. 1834.

Shell large, thick, heart-shaped, somewhat compressed ; beaks submedial, prominent, incurved, contiguous ; obsoletely radiately striate ; margin entire, gaping behind. Epidermis thin, pale olivaceous or drab, the young with occasionally zigzag darker lines ; within white or yellowish.

Length 2.7, height 2.3 inches.

Maine, northwards.

Family CHAMIDÆ.

Labial palpi small, curved, obliquely truncate. Mantle closed, margins united by a fringed curtain ; siphonal orifices small, wide apart, the branchial slightly prominent, with the orifice fimbriated, the anal with a simple valve ; gills two on each side, unequal, plicate. Foot cylindrical, bent. Living attached to stones and rocks.

Genus **CHAMA**, Linnæus.

Syst. Nat., edit. x. 1758.

1. C. ARCINELLA, Linnæus. Fig. 459, 462, 463.

Syst. Nat., edit. xii. 1139. 1767.

Shell heart-shaped, with a large depressed lunule beneath the umbones, both valves radiately ribbed, the ribs spinous, and the interstices punctured; margins very finely crenulated. White or yellowish, most frequently stained with pink-red.

North Carolina to West Indies.

The spines are generally only partially developed.

2. C. MACROPHYLLA, Chemnitz. Fig. 460.

Conch. Cab., vii. 149, t. 52, f. 514, 515. 1784.

Shell ovate, both valves lamellated throughout, lamellæ imbricated, large, irregular, striated; margins of the valves very minutely crenulated; color bright yellow, whitish within.

North Carolina to West Indies.

Family ARCADÆ.

Animal with the mantle open; foot large, bent, and deeply grooved; gills very oblique, united posteriorly to a membranous septum.

Synopsis of Genera.

Shell equivalve or nearly so, oval or subquadrate, ventricose, strongly ribbed or cancellated; margins smooth or dentated, close or sinuated ventrally; hinge straight, teeth very numerous, transverse; umbones anterior, separated by a flat, lozenge-shaped ligamental area, with numerous cartilage-grooves. ARCA, Linn.

Shell orbicular, nearly equilateral, smooth. or radiately striated; umbones central, divided by a striated ligamental area; hinge with a semicircular row of transverse teeth; margins crenate within.

PECTUNCULUS, Lam.

Shell trigonal, with the umbones turned towards the short *posterior* side; smooth or sculptured; epidermis olive, *interior pearly*, margins crenulated; hinge with a prominent internal cartilage-pit, and a series of sharp teeth on each side. NUCULA, Lam.

Shell oblong, rounded in front, produced and pointed behind; margin not crenated; *pallial line with a small sinus;* teeth as in *Nucula.*

LEDA, Schum.

12

Genus **ARCA**, Linnæus.

Syst. Nat., edit. x.   1758.

The animal has a long-pointed foot, deeply grooved, and heeled; mantle furnished with ocelli; palpi 0; gills long, narrow, less striated externally, continuous with the lips; hearts two, each with an auricle.

The Arcas with close valves have the left valve a little larger than the right, and somewhat overlapping at the margin.

There are about two hundred species; distribution universal, ranging from low-water to 230 fathoms.

1. A. NOÆ, Linnæus. Fig. 464.

Syst. Nat., edit. xii. 1140. 1767.

A. *zebra*, Swainson, Zool. Illust., No. 26, t. 118.

Shell elongately oblong, anterior side very short, posterior side emarginate, with a blunt keel extending from the umbone to the margin; ventral margin more or less gaping; white, with waved brownish streaks; radiately ribbed; ligament area flatly concave.

North Carolina to West Indies. (*Medit.*)

2. A. PONDEROSA, Say. Fig. 467.

Journ. Philad. Acad., ii. 267. 1822.

Shell very thick and ponderous, somewhat oblique, with 25 to 28 ribs, each marked with an impressed line. Beaks distant, opposite the middle of the hinge; lower margin nearly straight or even somewhat contracted in the middle.

Length 2.5, height 2 inches.

Southern Coast.

Fossil valves of this species sometimes occur on the beach at Cape May and Atlantic City, N. J.

3. A. TRANSVERSA, Say. Fig. 465.

Journ. Phil. Acad., ii. 269. 1822.

Shell transversely oblong, rhomboidal, with from 32 to 35 ribs, umbones separated by a long narrow area; extremities of the hinge margin angulated; epidermis chestnut-brown.

Length 30, height 8.5 mill.

New England, New York, southwards.

4. A. LIENOSA, Say. Fig. 469.

Am. Conch., iv. t. 36, f. 1. 1832.

This shell is described as fossil, and worn (probably fossil) valves are found abundantly at Beaufort, N. C. It is admitted here, because Dr. Stimpson has included it in his check-list of recent species, yet I suspect that it has not been found living.

5. A. PEXATA, Say. Fig. ~~466.~~ L 70
Journ. Philad. Acad., ii. 268. 1822.

Shell covered (when fresh) with a hairy epidermis, transversely subovate, with from 32 to 36 ribs, placed closer together than their own diameters; beaks far forward, near the anterior termination of the hinge approximate.

Length 57, height 43 mill.

Rhode Island, southwards.

6. A. AMERICANA, Gray. Fig. ~~470.~~ 466
Wood, Index Test. Suppl., t. 2, f. 1.

Shell ovately oblong, sides rounded, the anterior very short and contracted; white, covered with a rather thick blackish-brown epidermis, which is bristly in the interstices between the ribs; ribs about 35 in number, each one with a median impressed line, interstices deeply cut; ligament area very narrow; umbones anterior, nearly touching.

North Carolina to West Indies.

This shell is allied to *A. pexata*, but is somewhat larger, more oblong in shape, and the ribs are generally impressed. Mr. Reeve says (Conch. Icon.) that the ribs of *Americana* are flat, while those of *pexata* are impressed in the middle, but the contrary is the case in nearly all the specimens I have examined.

7. A. HOLMESII, Kurtz. Fig. 471.
Cat. Mar. Shells, 5. 1860.

Distinguished from the two preceding species by its smaller size, and more inflated and globular form; it is also more solid. Inhabits estuaries (Kurtz).

N. and S. Carolina.

8. A. INCONGRUA, Say. Fig. 472.
Journ. Philad. Acad., ii. 268, July, 1822.

Shell somewhat squarely orbicular, rather thin, very inequivalve, sides angulated at the upper part, anterior side the longest; white, with a thin light brown epidermis; ribs 27 or 28 in number, in the left valve the anterior ones are elevately crenated, in the right valve they are all crenated; ribs of the left value rather broader than those of the right; ligament area rather wide; umbones somewhat approximating.

Length 2+, height 2 inches.

N. Carolina to W. Indies.

Genus **PECTUNCULUS**, Lam.
Syst. 115. 1801.

Animal with a large crescent-shaped foot, margins of the sole undulated; mantle open; margins simple, with minute ocelli; gills equal, lips continuous with the gills.

About sixty species known, ranging from 8 to over 100 fathoms.

1. P. PENNACEUS, Lamarck. Fig. 473.
Anim. sans Vert.

P. lineatus, Reeve Zool. Proc. 1843.
P. spadiceus, Reeve, Zool. Proc. 1843.

Shell orbicular, swollen, decussately striated, longitudinal striæ the strongest; whitish, irregularly painted with large and small dark-brown spots and streaks; umbones bent inwards to the anterior end of the ligament.

N. Carolina to West Indies.

This shell has been doubtfully referred by some conchologists to *P. Charlestoniensis*, a post-pliocene fossil of S. Carolina.

Genus **NUCULA**, Lamarck.
Syst. 115. 1801.

Animal with the mantle open, its margins plain; foot large, deeply fissured in front, forming when expanded a disk with serrated margins; mouth and lips minute, palpi very large, rounded, strongly plaited inside, and furnished with a long convoluted appendage; gills small, plume-like, united behind the foot to the branchial septum.

Distribution about 70 species, from 5 to 100 fathoms.

1. N. TENUIS, Montagu. Fig. 478.
Test. Brit. Suppl. 56, t. 29, f. 1. 1808.

Shell small, thin, trapezoidal; smooth, without radiating lines; beaks prominent; epidermis grass-green; inner margin entire.

Length 7.5, height 6.25 mill.

Maine, northwards. (*Eur.*)

2. N. PROXIMA, Say. Figs. 479, 480.
Journ. Philad. Acad., ii 270. 1822.

Arca nucleus, Linn (part). Syst. Nat., edit. xii. 1143.

Shell oblique, ovate-triangular, crossed by minute concentric and radiating lines; epidermis olivaceous; margin crenulated; hinge-teeth large, twelve before and eighteen behind the beaks.

Length 11, height 9 mill.

Whole Coast, southwards to N. Car. (*Eur.*)

3. N. EXPANSA, Reeve. Figs. 481, 482.

> Belcher's Arctic Voy. 397, t. 33, f. 2. 1855.

N. *Bellotii*, Adams, Zool. Proc. 51. 1856.

Shell large, ovate-triangular, tumid, the surface distinctly nucleated with ridges, both dorsal areas with fine radiating striae; ten teeth in front and fifteen behind the beaks. Dark chestnut colored.

Length 14, height 9 mill.

Canadian waters, northward. (*Eur.*)

This may be only a very large flourishing state of *N. tenuis.*

4. N. INFLATA, Hancock. Figs. 483, 484.

> Ann. Nat. Hist. 333, t. 5, f. 13, 14. 1846.

Shell trapeziform, inflated, thin, coarsely concentrically striate; interior margin simple; hinge with five teeth before and ten behind the large oblique ligament cavity; epidermis yellowish-green.

Length 7.5, height 6.25 mill.

Labrador, northwards.

5. N. DELPHINODONTA, Mighels. Figs. 485–487.

> Bost. Journ. Nat. Hist., iv. 40, t. 4, f. 5. 1842.

Shell minute, obliquely triangular, beaks raised, nearly terminal; hinge with three anterior and seven posterior teeth; epidermis olivaceous.

Length 3.25, height 2.75 mill.

New England, northwards.

Undetermined Species.

N. RADIATA, De Kay. Moll. N. York 179, t. 12, f. 216. 1843.

Dredged in E. River, N. Y. The figure is not recognizable.

Genus LEDA, Schumacher.
Essai Nov. Syst. 1817.

The animal is furnished with two partially united, slender, unequal siphonal tubes; gills narrow, plume-like, deeply laminated, attached throughout; mantle-margin with small ventral lobes, forming by their apposition a third siphon.

The typical group comprises about 80 species, inhabiting Arctic and northern seas, 10 to 180 fathoms.

1. L. TENUISULCATA, Couthouy. Figs. 488, 489.

> (*Nucula.*) Bost. Journ. Nat. Hist., ii. 64, t. 3, f. 8. 1838.

Nucula minuta, Gould, Invert. Mass., edit. i. 101. 1841.

Shell ovate-lanceolate, produced, narrowed and rostrated behind, covered with numerous concentric ridges; epidermis greenish-yellow to brownish; teeth twelve anterior and sixteen posterior to the beaks.

Length 25, height 11 mill.

New England, northwards.

2. L. JACKSONII, Gould.   Figs. 490, 491.
   (*Nucula.*)   Invert. Mass., edit. i. 102, f. 5.   1841.
Leda buccata, Steenstrup, Moller, Moll. Grœnl. 17.   1842.

Shell ovate, swollen, a little beaked and narrowed behind, surface concentrically ridged; teeth 15 in front and 20 behind the beaks.

Length 25, height 14 mill.

Maine, northwards.

This shell is higher in proportion to its length, and not so narrowly rostrated as *L. tenuisulcata;* it differs also in the number and arrangement of the teeth.

3. L. MINUTA, Fabricius.   Figs. 492, 493.
   (*Arca.*)   Fauna Grœnl. 414.   1780.

Shell oblong, inflated, somewhat pear-shaped, posterior side not much produced; brownish; 12 teeth before and about 14 behind the beaks.

Length 12.5, height 7.5 mill.

Halifax, N. S.

More nearly equilateral than the other species.

4. L. CAUDATA, Donov.   Figs. 494, 495.
   (*Arca.*)   Brit. Shells, t. 78.

Shell long, depressed, slender; epidermis yellowish, ridged.

Length 15, height 6.25 mill.

Halifax, northwards.

Smaller than *tenuisulcata*, and more recurved posteriorly; the beaks also are more acute and less tumid.

5. L. ACUTA, Conrad.   Fig. 496.
   (*Nucula.*)   Am. Mar. Conch., t. 6, f. 3.

Shell ovate, elongated, convex, with numerous, regular, concentric striæ; posterior side slightly recurved, and very acute at the extremity; epidermis dark green.

Length 6, height 4 mill.

North Carolina.

This species was first described by Mr. Conrad as a doubtful fossil; it lives abundantly near Fort Macon, N. C.

6. L. UNCA, Gould.

Bost. Proc., viii. 280. 1862.

Shell small, reddish, subequilateral, covered with profound ridges; anteriorly broadly rounded ; posteriorly very acute; posterior dorsal margin concave, cristate, smooth; posterior ventral margin sub-emarginate ; teeth 12 to 15.

Length 8, height 6 mill.

Frying-pan Shoals, N. C.

I have not seen this species, and it has never been figured. From the description, I think it possibly identical with *acuta*, Conr.

Subgenus YOLDIA, Müller.

Kröyer's Naturh. Tidsskr., iv. 91. 1832.

Shell oblong, slightly attenuated behind, compressed, thin, smooth, or obliquely sculptured, with a polished epidermis; pallial sinus deep.

1. L. LIMATULA, Say. Fig. 497.

(*Nucula.*) Am. Conch., t. 12, 1831.

Shell ovate-oblong, slightly rostrated posteriorly; epidermis light green ; beaks not prominent, subcentral; with 22 anterior and 18 posterior teeth.

Animal with united siphons, the anal one translucent, the branchial opaque white, both fringed at the openings; edges of foot lobes simple.

Length 57.5, height 27.5 mill.

New England (northwards), N. Car.

The above dimensions, from specimens dredged in Portland Harbor, Maine, are about double the size of ordinary specimens. The animal is said to be very active, leaping to an astonishing height.

2. L. SAPOTILLA, Gould. Fig. 498.

(*Nucula.*) Invert. Mass. 100, f. 61. 1841.

Shell ovate-oblong, very slightly rostrated posteriorly ; epidermis light yellowish-green ; beaks subcentral, tumid, with a slight flexure under the posterior tip; teeth sixteen or eighteen on each side.

Length 22.5, height 10 mill.

New England, northwards.

Probably only a variety of *L. limatula.*

3. L. MYALIS, Couthouy. Fig. 499.
(*Nucula.*) Bost. Journ. Nat. Hist., ii. 61, t. 3, f. 7. 1838.

Shell ovate, smooth, olive-colored; anterior part longest and rounded; posteriorly acuminated and sub-rostrated; teeth about 12 on each side, sometimes increased to 16 or 18 in number.

Length 27.5, height 17.5 mill.

New England, northwards.

This shell is higher in proportion to its width, and darker colored than L. *limatula;* it also differs in the position of the beaks and number of teeth.

4. L. OBESA, Stimpson. Figs. 500, 501.
Proc. Bost. Nat. Hist., iv. 13. 1851.

Shell small, thin, inflated, oval, smooth; beaks nearly central; teeth small, 10 in front and 12 behind; epidermis pale yellowish-green.

Length 6, height 4 mill.

(Deep-water.) *Massachusetts Bay.*

Closely allied to *Yoldia pygmæa,* Münst., but that shell is more pointed and upturned at the posterior end.

5. L. ARCTICA, Gray. Figs. 502, 503.
Nucula glacialis, Gray, Wood Index Test. Suppl., t. 2, f. 6.
Nucula truncata, Brown, Brit. Conch. 84, t. 33, f. 18.
Nucula Portlandica, Hitchcock, Bost. Journ., i. 327.
Nucula siliqua, Reeve, Belcher's Arctic Voy., t. 33, f. 4. 1855.

Shell oblong, ovate, ventricose; beaks prominent, nearly median; posteriorly sulcate between two slight rounded elevations from beak to margin; 12 to 14 teeth each side of the ligamental spoon.

Length 20, height 12.5 mill.

(*Semi-fossil in the clays at Portland.*) *Arctic Seas.*

6. L. THRACIÆFORMIS, Storer. Fig. 504.
(*Nucula.*) Bost. Journ. Nat. Hist., ii. 122.
Nucula navicularis, Couthouy, Bost. Journ., ii. 178, t. 4, f. 4 (young).

Shell subquadrate, rounded in front, truncately rounded behind; beaks anterior, with a slight elevation or rib proceeding to basal posterior margin; epidermis dusky-green, lighter posteriorly; hinge with 12 teeth each side of the spoon-shaped cavity.

Length 70, height 40 mill.

New England? Arctic. (*From fishes.*)

7. L. CASCOENSIS, Mighels.   Figs. 505–7.
(*Nucula.*) Bost. Journ. Nat. Hist., 40, t. 4, f. 6.

Shell ovate, rather thin, finely striate, slightly inequilateral; anterior side semi-oval; posterior side tapering nearly to a point, with a well-defined areola, sharply compressed, with a slight wave below the areola; epidermis greenish straw-color; beaks small, nearly central; teeth small, 10 anterior and 10 or 12 posterior.

Length 15, height 9 mill.

Casco Bay, Maine.

Family MYTILIDÆ.

Animal marine (or sometimes fluviatile), attached by a byssus; mantle-lobes united between the siphonal openings; gills two on each side, elongated, and united behind to each other and to the mantle; dorsal margins of the outer and innermost laminæ free; foot cylindrical, grooved.

Synopsis of Genera.

Shell wedge-shaped, rounded behind, *umbones terminal*, pointed; hinge-teeth minute or obsolete; pedal muscular impressions *two* in each valve, small, simple, close to the adductors.   MYTILUS, Linn.

Shell *oblong*, inflated in front; *umbones anterior*, obtuse; hinge toothless; pedal impressions *three* in each valve, the central elongated; epidermis sometimes produced into long beard-like fringes.   MODIOLA, Lam.

Shell *cylindrical*, interior nacreous; otherwise like *Modiola*.
    LITHODOMUS, Cuv.

Shell short, ovate, partly smooth, and partly ornamented with radiating striæ; hinge margin crenulated behind the ligament; interior brilliantly nacreous.   CRENELLA, Brown.

Shell ovate, oblong, obtusely keeled right valve with a slight byssal sinus; beaks terminal, furnished internally with a transverse shelf or septum; hinge composed of an imperfectly developed cardinal tooth in the right valve, with a corresponding socket in the left; ligament linear, internal; pedal impression single, posterior.   Fluviatile.
    DREISSENA, Van Beneden.

Genus MYTILUS, Linnæus.
Syst. Nat., edit. x., 705. 1758.

The common edible mussel frequents mud-banks which are uncovered at low water; the fry abound in water a few fathoms deep; they are full-grown in a single year. Pieces of wreck are frequently covered by mussels of all ages, attached by their byssus. There are about sixty-five species, of world-wide distribution.

1. M. EDULIS, Linnæus. Figs. 474–477.
Syst. Nat., edit. xii., 1157. 1757.

M. pellucidus, Pennant, Brit. Zool., iv. 237, t. 66, f. 3. 1777.
M. notatus, DeKay, Nat. Hist., N. York, 182, t. 13, f. 223. 1843.

Shell ovate, oblong, beaks pointed, basal margin nearly straight, ligament margin straight; posteriorly widened and rounded; hinge with a few denticulations; epidermis dark-bluish, smooth, violet beneath; within white, with a broad blue margin.

VAR. *pellucidus*, shell horn-color, with blue rays, or uniform horn-color.

Length 60, height 32.5 mill.

Whole Coast. (Eur.)

2. M. EXUSTUS, Linnæus. Fig. 508.
Mus. Ulric., 540.

M. cubitus, Say, Journ. Acad. Nat. Sci., ii. 263. 1822.

Shell oblong, striated, with elevated subglabrous lines, which are smaller on the anterior side. Yellowish, sometimes blotched with green or brownish.

Length 28, height 12.5 mill.

Southern Coast, West Indies.

3. M. HAMATUS, Say. Fig. 509.
Journ. Philad. Acad., ii. 264. 1822.

M. striatus, Barnes, Am. Journ. Sci., vi. 364. 1823.

Shell very much contracted and incurved on the basal margin; valves very much striated on every part with radiating elevated lines, which sometimes become bifid or trifid; color dark fuscous; within purplish.

Length 28, height 20 mill.

Southern Coast, W. Indies.

Genus MODIOLA, Lamarck.
Syst., 113. 1801.

1. M. MODIOLUS, Linnæus. Figs. 510, 511.
(*Mytilus.*) Syst. Nat., edit. xii., 1158.

M. papuana, Lam., Anim. s. Vert., vii. 17.
Mytilus barbatus, Linn., Syst. Nat., xii. 1158 (young).

Shell large, coarse, and solid, oblong, obliquely dilated; beaks tumid, obtusely angulated; basal margin concave, with a fissure for the byssus; surface coarsely marked by growth-lines; epidermis thick, dark violaceous, blackish, or chestnut-brown; within pearly.

Length 4.5–6, height 2.5–3 inches.

Animal dark-orange or reddish; edible.

Northern Coast, northwards. (*Eur.*)

2. M. PLICATULA, Lamarck. Fig. 512.

Anim. s. Vert., vii. 22. 1822.

M. semicostata, Conr., Journ. Phila. Acad., vii. 244, t. 20, f. 7.

Shell oblong, obliquely dilated, somewhat falciform; surface with approximated deep furrows radiating towards the dilated margin, fainter on the basal margin, but more distinct near the beaks, which are often eroded; a few distant, concentric, narrow, impressed lines cross the radiating striæ; beaks prominent, rounded; hinge margin straight, ascending; basal margin concave, depressed, with a small fissure for the byssus. Epidermis greenish-yellow to reddish-brown; within pearly, occasionally purple-tinted.

Length 2.5–4.5, height .8–1.5 inches.

Whole Coast.

Inhabits salt marshes, estuaries, and brackish waters.

3. M. TULIPA, Linnæus. Fig. 514.

Modiola Americana, Leach., Zool. Misc., ii. t. 72, f. 1. 1815.
Modiola castanea, Say, Journ. Philad. Acad., ii. 266. 1822.

Shell oblong, rather thin, ventricose; hinge margin elevated in a right line from the beak to the alated angle, from which it declines in a right line to nearly an equal distance, the alar angle rounded; anterior margin short and small; basal margin slightly contracted in the middle. Epidermis marked only by growth lines, yellowish or brownish, with dark rays over the middle posterior portion, sometimes uniform chestnut-color.

Southern Coast. (*W. Ind.*)

M. castanea appears in some catalogues as a distinct species, but an author's specimen? in Mus. Philad. Acad. proves its identity with *tulipa*.

4. M. CAROLINENSIS, Conrad. Fig. 513.

Journ. Philad. Acad., vii. 244, t. 20, f. 6. 1837.

Shell dilated in the middle; disks with very numerous radiating striæ; lower margin rounded and beautifully crenulate. Color greenish-yellow; within yellowish, spotted with purple.

Charleston, S. C.

The figure is a copy of the original. I have not been able to identify this species.

Genus **LITHODOMUS**, Cuvier.
Reg. Anim., ii. 461.   1817.

The animal, which is eaten in the Mediterranean, is like a common mussel; but differs in habit, boring into corals, shells, and the hardest limestone rocks; its burrows are shaped like the shell, and do not admit of free rotatory motion. The genus inhabits warm seas.

1. L. FORFICATUS, Ravenel.
   Proc. Philad. Acad., 44.   1861.

Shell thin, fragile, white; posterior end with a narrow projection on each valve, deflected so as to cross each other; within light salmon color.
Length 31 mill.

Charleston, S. C.

From a mass of coral drawn up by a fishing line, in 14 fathoms off Charleston Bar. There was quite a colony of these shells in the coral. Possibly ballast from some distant locality? A similar species inhabits the Caribbean Sea.

Genus **CRENELLA**, Brown.
Hist. Brit. Conch. 1827.

There are about 25 species of this genus, inhabiting temperate and arctic seas. Low water to 40 fathoms. Spinning a nest, or hiding amongst the roots of sea-weed and corallines.

a. Typical species. Surface of valves entirely covered by striæ, radiating in two diverging fasciculi from the beaks. Shell suborbicular or oval.

1. C. GLANDULA, Totten.   Fig. 515.
   (*Modiola.*) Am. Journ. Sci., xxvi. 367, f. 3, c. f. g.
Mytilus decussatus, Stimpson, Shells N. E., ii. 1851.

Shell oblique, oval, orbicular, inflated, thin, radiating lines crowded; inner margin crenulated; epidermis brownish-yellow; within pearly.
Length 12, breadth 9 mill.

New England, northwards.

2. C. FABA, Fabricius. Fig. 516.

(*Mytilus.*) Fauna Grœnl. 1780.

Modiola pectinula, Gould, Invert. Mass., 1st edit. 127, t. 85.

Shell obovate, ventricose, with about forty equal radiating ribs; beaks prominent, projecting as far as the anterior margin; margin crenulated within; epidermis brownish-yellow.

Length 17.5, breadth 12.5 mill.

Arctic Seas to Greenland.

b. Sides of the shell with radiating lines, middle portion smooth. S. G. MODIOLARIA, Gray.

3. C. NIGRA, King. Figs. 517, 518, 519.

Ann. Nat. Hist., xviii. 239.

Mytilus discrepans (?) Mont., Test. Brit. Suppl., 65, t. 26, f. 4.

Modiola nexa, Gould, Invert. Mass., 1st edit. 128, f. 86

Shell ovate; beaks prominent, and placed considerably behind the anterior extremity; epidermis rusty-brown.

Length 2.5, breadth 1.5 inches.

Massachusetts, northwards.

The young shell (Fig. 518), represents *M. nexa*, Gould.

4. C. DISCORS, Linnæus. Figs. 520, 521.

(*Mytilus.*) Syst. Nat., edit. xii. 1159. 1767.

Mytilus discrepans, Mont., Test. Brit., 169. 1803.

Modiola lævigata, Gray, App. Parry's 2d Voyage, 245. 1824.

Shell suboval, broadest behind; beaks nearly terminal; hinder extremity somewhat lobed; anterior ribs about eight; posterior ones numerous; greenish-yellow, with clouds of olive, becoming nearly black in old specimens.

Length 37, breadth 20 mill.

Massachusetts, northwards. (*Eur.*)

The smaller figure represents *discors* (not fully grown), and the large figure is from an adult specimen (Gray's *lævigata*).

5. C. CORRUGATA, Stimpson. Fig. 522.

(*Mytilus.*) Shells N. E., 12. 1851.

Modiola discors, Gould, Invert. Mass., 130, f. 84 (not of English authors.)

Shell oval, tumid; upper edge somewhat compressed and arching; posterior tip somewhat produced and pointed; beaks large, nearly terminal; surface with about sixteen ribs at the anterior third, and very numerous ones at the posterior third; three or four teeth before the beaks; epidermis greenish-yellow.

Massachusetts, northwards.

This species will probably prove to be a mere variety of the following.

6. C. LATERALIS, Say.   Fig. 523.

(*Mytilus.*)   Journ. Philad. Acad., ii. 264. 1822.

Shell transversely suboval, inflated, subpellucid, with numerous concentric wrinkles; ribs alternately larger and smaller; shell inflated from the beak to the posterior basal angle; epidermis greenish or brownish.

Southern Coast.

Genus **DREISSENA**, Van Beneden.
Bull. Brux. Acad., 25. 1835.

Animal with closed mantle; byssal orifice small, and siphon very small, conical, plain, branchial prominent, fringed inside; palpi small, triangular.

Inhabits brackish or fresh waters.

1. D. LEUCOPHÆTA, Conrad.   Fig. 524.

(*Mytilus.*)   Journ. Philad. Acad., vi. 263, t. 11, f. 13. 1831.

Shell incurved, with a very rugose, brownish epidermis; anterior side much depressed.   Hinge.margins excavated, with the teeth obsolete.

Chesapeake Bay, southwards.   (Brackish water.)

Family AVICULIDÆ.

Animal with the mantle-lobes free, their margins fringed; foot small, spinning a byssus; gills two on each side, crescent-shaped, entirely free, or united to each other posteriorly or to the mantle.

These shells are natives of tropical and warm seas; no living species are found in northern latitudes.

Synopsis of Genera.

Shell obliquely oval, very inequivalve; right valve with a byssal sinus beneath the anterior ear; cartilage pit single, oblique; hinge with one or two small cardinal teeth, and an elongated posterior tooth, often obsolete; posterior muscular impression (adductor and pedal) large, sub-central; anterior (pedal) scar small, umbonal.   AVICULA, Brug.

Shell equivalve, wedge-shaped; umbones quite anterior; posterior side truncated and gaping; ligamental groove linear, elongated; hinge edentulous; anterior adductor scar apical, posterior sub-central, large, ill-defined; pedal scar in front of posterior adductor.   PINNA, Linn.

Genus **AVICULA**, Brug.
Encyc. Meth., t. 177. 1792.

1. A. ATLANTICA, Lamarck. Fig. 525.
Anim. s. Vert., vii. 1822.
Avicula hirundo, Say. Journ. Acad. Nat. Sci. Philad., ii. 262. 1822.

Shell reddish-brown, with very numerous undulated wrinkles, which are disposed in radii and rendered more conspicuous by a white longitudinal line at the junction of each series of wrinkles with its contiguous one.

North Carolina, southwards.

Genus **PINNA**, Linnæus.
Syst. Nat., edit. x. 707. 1758.

Animal with the mantle doubly fringed; foot elongated, grooved, spinning a powerful byssus attached by large triple muscles to the centre of each valve; adductors both large; palpi elongated; gills long.

1. P. SEMINUDA, Lamarck. Fig. 526.
Anim. sans Vert., vii. 61. 1822.
P. squamosissima, Philippi.

Shell triangular, truncated; posterior side longitudinally ribbed, many scaled, scales delicate, erect; anterior side with the scales plentiful, minute, rather obsolete towards the umbones, with a few large, concentric wrinkles, dull olive.

South Carolina, southwards.

2. P. MURICATA, Linnæus. Fig. 527.
Syst. Nat., edit. xii. 1160. 1767.
P. Carolinensis, Hanley. Proc. Zool. Soc. 225. 1858.

Shell triangular, whitish, somewhat ventricose, obscurely ribbed, ribs armed with triangular, erect scales.

South Carolina, southwards.

Family OSTRÆIDÆ.

Synopsis of Genera.

Shell irregular, attached by the left valve; upper valve flat or concave, often plain; lower valve convex, often plaited or foliaceous, and with a prominent beak; ligamental cavity triangular or elongated; hinge toothless; structure subnacreous, laminated. OSTREA, Linnæus.

Shell orbicular, very variable, translucent, and slightly pearly within, attached by a plug passing through a hole or notch in the right valve ; upper valve convex, smooth, lamellar or striated ; interior with a submarginal cartilage-pit and four muscular impressions, three subcentral, and one in front of the cartilage ; lower valve concave, with a deep rounded notch in front of the cartilage process ; disk with a single (adductor) impression.   ANOMIA, Linnæus.

Shell suborbicular, regular, resting on the right valve, usually ornamented with radiating ribs ; beaks approximate, eared ; anterior ears most prominent ; posterior side a little oblique ; right valve most convex, with a notch below the front ear ; hinge-margins straight, united by a narrow ligament ; cartilage internal, in a central pit ; adductor impression double, obscure ; pedal impression only in the left valve, or obsolete.
                                    PECTEN, Müller.

Shell equivalve, compressed, obliquely oval ; anterior side straight, gaping, posterior rounded, usually close ; umbones apart, eared ; valves smooth, punctate-striate or radiately ribbed and imbricated ; hinge-area triangular, cartilage-pit central ; adductor impression lateral, large, double ; pedal scars two, small.   LIMA, Brug.

Shell irregular, attached by the umbo of the right valve ; valve smooth or plaited ; hinge-area obscure ; cartilage quite internal ; hinge-teeth two in each valve ; adductor scar simple.   PLICATULA, Lam.

Genus **OSTREA**, Linnæus.
Syst. Nat., edit. x. 696.   1758.

Animal with the mantle-margin double and finely fringed ; the gills are nearly equal, united posteriorly to each other and the mantle-lobes, forming a complete branchial chamber ; lips plain ; palpi triangular, attached ; sexes distinct.

1. O. VIRGINIANA, Lister.   Figs. 528, 529.
     Conch., t. 200, f. 34.   1686.
Ostrea Virginica, Gmel. Syst. Nat., 3336.   1790.
Ostrea Canadensis, Lam. Anim. sans Vert., vii. 226.   1822.

Shell narrowly elongated, whitish, thick-lamellar ; upper valve rather plane ; becoming thick with age, the lower beak projecting and with an inner channel transversely channelled ; muscular impression chestnut or violet-color.

Varies from 6–12 inches in length, and 3–4 inches in breadth.
                                    Whole Coast.

This is the common edible oyster of Chesapeake bay ; it is native about as far north as New York, where it is replaced by the northern species O. *borealis.* It is also found on the New

England coast and Gulf of St. Lawrence, but has probably been transplanted to these localities. The lower valve is sometimes ornamented with red or violet rays.

2. O. BOREALIS, Lamarck. Figs. 530, 531.
   Anim. sans Vert., vii. 220. 1822.

Shell rounded-ovate, the upper valve covered with membranous scales, the lower valve irregularly spiny-ribbed and foliaceous.

Length 3 to 6 inches, breadth about 2 to 5 inches.

New England, New York.

This species is apparently very distinct from the preceding; it is smaller, wider, has not the lengthened beak of the lower valve of *Virginiana* and the surface is much rougher. It is *very* closely allied to *O. edulis* of Europe.

3. O. TRIANGULARIS, Holmes.
   Proc. Elliott Soc. Nat. Hist., 29. 1856.

Shell subtriangular, subequivalve, subequilateral, thick, laminated; beaks produced, acutely pointed, angular, and slightly curved towards each other; margins rounded; cavity of the shell circular; muscular impression very large in proportion to the size of the shell, and placed near the margin of the base.

South Carolina.

Dr. Holmes states that this shell resembles *O. edulis* of Europe but is more regular in form. Its large muscular impression, pointed beaks, and triangular shape distinguish it from that species. I am not acquainted with this shell.

4. O. EQUESTRIS, Say. Figs. 532, 533.
   Am. Conch., vi. t. 58. 1834.

Shell small, ovate-triangular, more or less folded longitudinally; lateral margins near the hinge with from six to twelve denticulations of the superior valve received into corresponding cavities of the inferior valve; superior valve depressed, but slightly folded; inferior valve convex, attached by a portion of its surface, the margins elevated, folds unequal, much more profound than those of the superior valve; hinge very narrow, and curved laterally and abruptly.

North Carolina to West Indies.

13

Doubtful Species.

O. semicylindrica, Say. This appears to be an immature shell and is not identified. It is said to inhabit the coast of Georgia and Florida, imbedded in sponges.

Genus **ANOMIA**, Linnæus.
Syst. Nat., edit. xii. 1150. 1767.

Animal with the mantle open, its margins with a short double fringe; lips membranous, elongated; palpi fixed, striated on both sides; gills two on each side, united posteriorly, the outer laminæ incomplete and free; foot small, cylindrical, subsidiary to a lamellar and more or less calcified byssal plug, attached to the upper valve by three muscles; adductor muscle behind the byssal muscles, small, composed of two elements; sexes distinct; ovary extending into the substance of the lower mantle-lobe.

There are about twenty species; distribution principally in temperate seas, from low water to 100 fathoms.

1. A. GLABRA, Verrill. Fig. 534.
   Am. Journ. Sci., 288. April. 1872.
A. ephippium, Gould. Invert. Mass., edit. i. 1841.
A. electrica, Gould. Invert., edit. ii. 205. 1870.

Shell orbicular, or distorted; surface scaly, lamellar, and easily impressed by contact with other shells, etc.; upper valve very convex, with a small beak; lower valve smaller, flat, or concave, with a circular byssal hole, which is united to the margin by a greater or less fissure. Polished, and varying in color from black through red, yellow, and ash to white; the same colors internally, except that the muscular impression is opaque white.

Diameter usually about 1 inch.

Cape Cod, Mass. to Florida.

This is our common *Anomia*, generally known as *A. ephippium*, but it appears to be distinct from the European shell bearing that name.

2. A. ACULEATA, Gmelin. Figs. 535, 536.
   Syst. Nat., 3346. 1790.

Shell small, rounded; upper valve with fine, prickly scales arranged in radiating lines; lower valve smooth; yellowish or whitish.

Diameter about half an inch.

Eastport, Maine, northwards.

A. SQUAMULA, Linn. Gould, Invert. Mass., edit. 2, 206. 1870.
I do not know this species; it does not appear, from the description, to differ from *A. glabra*.

Genus PECTEN, O. F. Müller.
Zool. Dan. Prodr., p. xxxi. 1776.

Animal with the mantle quite open, its margins double, the inner pendant like a curtain finely fringed ; at its base a row of conspicuous round, black eyes, surrounded by tentacular filaments; gills exceedingly delicate, crescent-shaped, quite disconnected posteriorly, having separate excurrent canals ; lips foliaceous; palpi-truncated, plain outside, striated within ; foot fingerlike, grooved, byssiferous in the young.

Unlike the oyster, this is an active animal, having the power of rapid motion. There are nearly two hundred species; of worldwide distribution, extending to 200 fathoms in depth.

1. P. MAGELLANICUS, Gmelin. Fig. 537.
(*Ostrea*). Syst. Nat., 3317. 1790.
Pecten tenuicostatus, Mighels & Adams. Bost. Journ. Nat. Hist., iv. 41, t. 4, f. 7. Bost. Proc., i. 49. 1841.
Pecten fuscus, Linsley. Am. Journ. Sci., 1st ser., xlviii. 278. 1845 (young).
Pecten brunneus, Stimpson. Shells N. E., 8. 1851 (young).

Shell large, orbicular, inequivalve ; superior valve more convex, dull-red, with very numerous radiating striæ which are crossed by minute subscabrous wrinkles; inferior valve nearly flat, whitish, with the striæ less distinct ; beaks purple ; white within.

Diameter 6–9 inches.

New England, New Jersey, West Indies?

This is our largest species, and one of the largest in the genus. It is distinguished by the absence of ribs, the surface being only marked by fine striæ. Most American authors have adopted Mighels' name because the species does not come from the Straits of Magellan as supposed by Gmelin ; but this is mere supposition, and I think it is preferable to retain the original name. It has been considered a northern species, but in the collection of the Philadelphia Academy are fine specimens dredged alive in Raritan Bay, also specimens from "West Indies."

2. P. ISLANDICUS, Müller.   Fig. 538.
        (*Ostrea.*) Zool. Danica, Prod. No. 2990.   1776.
Pecten Pealii, Conrad. Am. Mar. Conch., t. 12, f. 2.   1831.

Shell oblong, orbicular, slightly oblique, valves nearly equal. Surface covered with numerous, small, scaly, radiating ribs; ears unequal in size. Valves closed except at the notch. Color light-orange to dark reddish-brown, frequently zoned or blotched on the upper valve, lower valve lighter in color; white within, except a large roseate spot near the beaks.

Length 3.5 inches.

New England. (Eur.)

3. P. IRRADIANS, Lamarck.   Fig. 539.
        Anim. sans Vert., vi. Pecten No. 37.   1819.
P. concentricus, Say, Journ. Philad. Acad., ii. 259.   1822.

Shell orbicular, with from eighteen to twenty elevated, rounded ribs, and numerous concentric wrinkles; inferior valve slightly ventricose or gibbous towards the umbo; ears large and nearly equal. The upper valve is generally brown with pale zones, and the lower valve yellowish or whitish with pale brown zones.

Diameter 2–3 inches.

Whole Coast.

The animal of the "scallop" is eaten, and is comparable in flavor to the lobster. The species is particularly numerous and well-grown on the shores of Long Island and New Jersey, where the young shells, on a clear calm day, may be observed skipping along to a considerable distance on the surface of the water; the movement being accompanied by the noise occasioned by the rapid closing of their valves. The adults are not active.

4. P. DISLOCATUS, Say.   Fig. 540.
        Journ. Philad. Acad., ii. 260.   1824.

Shell suborbicular, with twenty or twenty-two rounded ribs, and very numerous concentric wrinkles; longitudinal striæ none; whitish tinged with purple or yellow, with a few narrow, trans-verse, interrupted and dislocated sanguineous, undulated lines, and sometimes five or six pale-reddish, almost obsolete spots towards the margin; ears subequal.

Diameter 1.5 inches.

North Carolina, southwards.

5. P. ORNATUS, Lamarck.　Fig. 541.
　　Anim. sans Vert., vi. 176.　1819.

Subequivalve, depressed, inequilateral, oblique, ribs thirty to
thirty-six, alternately smaller and subscabrous; one ear minute,
yellow ochraceous or white profusely blotched with angular red
markings.

Length 1 inch, width 20 mill.
　　　　　　　　　　　　North Carolina, southwards.

6. P. NODOSUS, Linnæus.　Fig. 542.
　　(*Ostrea.*)　Syst. Nat., edit. xii. 1145.　1767.

Shell with nine thick rounded ribs, and strong radiating striæ;
ribs with large hollow vesicles.　Reddish-brown, orange, or white.

Diameter 2-5 inches.
　　　　　　　　　　　　North Carolina, southwards.

P. PUSTULOSUS, Verrill, Amer. Journ. Sci., v. 14.　1873.　"St.
George's Banks."

This is probably, from its small size, an immature shell.　It has
not been figured, and besides, is extra-limital.

Genus LIMA, Bruguiere.
Encyc. Meth., t. 20, f. 6.　1792.

The animal has double mantle-margins, the inner pendent,
fringed with long tentacular filaments, ocelli inconspicuous; foot
finger-like, grooved; lips with tentacular filaments, palpi small,
striated inside; gills equal on each side, distinct.

The Limas are either free or spin a byssus; some make an
artificial burrow when adult, by spinning together sand or coral
fragments and shells.　The valves are always white.　About
twenty species are known.

1. L. SCABRA, Dillwyn.　Fig. 543.
　　Recent Shells, 271.　1815.

Ostrea glacialis, Gmelin (pars), Syst. Nat.

Shell oval, subequilateral, with numerous subscabrous striæ;
margin entire.

Length 2.25, width 1.5 inches.
　　　　　　　　　　　　Southern Coast.　West Indies.

2. L. SQUAMOSA, Lamarck.　Fig. 544.
　　Anam. sans Vert., vi. sp. 2.　1819.

Shell ovate, inequilateral, with strong, scaly ribs; hinge oblique;
margin plicated. ·

Length 1.5, width 1 inch.
　　　　　　　　　　　　Southern Coast to West Indies.

3. **L. SULCULUS**, Leach.   Figs. 545, 546.
    Forbes and Hanley, Brit. Moll., t. 53, f. 4, 5.

This beautiful little European species has been found at Sable Island, and will possibly be detected in our northern waters.

Genus **PLICATULA**, Lamarck.
Syst. An., 132.   1801.

1. P. RAMOSA, Lamarck.   Fig. 547.
    Anim. sans Vert., vi. 6.   1819.

Shell oblong-triangular, very stout and solid, with numerous large, ramified folds.   White, with ferruginous markings.

Length 1 inch.

North Carolina, southwards.

CLASS BRACHIOPODA.

Animals provided with a shelly covering composed of two valves, each of which is bilaterally symmetrical, and to which it is organically attached by three principal pairs of muscles. Soft parts also bilaterally symmetrical, consisting essentially of a mantle composed of two lobes, to which the valves correspond, of which lobes the outer edges are disunited throughout the greater part, or the whole of their extent; a disk and membrane, variously modified in form, with its edges fringed with a series of tubular brachia; the mouth situated within this disk at its posterior portion. Respiration performed by direct contact of sea-water with the vascular tissues of the brachia and mantle lobes; diœcious in all the genera? reproducing by ova only.

Family TEREBRATULIDÆ.

Shell rounded or oval; larger valve prominently beaked, smaller one provided internally with a shelly loop to which the brachia are attached. Valves articulated by two teeth in one valve received into sockets in the other.

Genus **TEREBRATULINA**, Orb.
Comp. Rend., xxv. 268. 1847.

1. T. SEPTENTRIONALIS, Couthouy. Figs. 548, 549.

(*Terebratula.*) Bost. Journ. Nat. Hist., ii. 65, t. 3, f. 18.

Terebratula caput-serpentis, Gould (non Auct.), Invert. Mass., edit. i. 141.
1841.

Shell obovate, whitish or yellowish-white, thin, translucent; upper
valve truncated horizontally at the apex; foramen large, one side
completed by the apex of the lower valve; surface covered by
minute radiating striæ.

Length 15, width 12 mill.

Maine, northwards.

Family RHYNCHONELLIDÆ.

Shell with radiating ribs, the arm supports long, slender, simple,
and gently curving towards each other; no area; the opening for
the pedicle usually completed by two small pieces; animal with
elongated spiral arms.

Genus **RHYNCHONELLA**, Fischer.
Mem. Soc. Imp., Moscow, ii. 1809.

1. R. PSITTACEA, Gmelin. Fig. 550.

(*Anomia.*) Syst. Nat., 3348. 1790.

Shell subtriangular, thin, inflated, brownish or greenish; beak
produced and curved; surface radiately finely striated. The
internal processes consist of two slender curved parallel prongs
proceeding from the base of the teeth of the upper valves.

Length 12, width 8 mill.

Newfoundland, northwards.

Family LINGULIDÆ.

Shell pedunculate, peduncle passing between the valves; inar-
ticulate, subequivalve; brachia unsupported by calcified processes.

Genus **GLOTTIDIA**, Dall.
Am. Journ. Conch., vi. 157. 1871.

Shell linguiform, smooth, elongated; neural valve furnished
internally with two sharp narrow incurved laminæ, diverging from

the beak and extending about one-third the length of the shell; hæmal valve with a mesial septum of about the same length extending forward from the beak.

1. G. PYRAMIDATA, Stimpson.   Fig. 551.

    (*Lingula*) Am. Journ. Sci., xxix. 444. 1860.

Shell greenish-white, ovate, elongated, convex; base subtruncate; surface smooth and glossy ; incremental lines inconspicuous.

Length 22, width 9 mill.

North Carolina.

The mantle has well-developed marginal setæ, those on either side, at the extremity longer than the rest, equalling in length one-third the width of the shell.   There are two black spots on the margin of the mantle at the extremity.   Peduncle, thick, muscular, hyaline with an opaque axis, three times the length of the shell.

This species has been found, so far, unattached. It is extremely active in its motions when disturbed, and has the power, as described by Mr. Morse, of burrowing in and travelling over the sand by contortions of the peduncle and movements of the setæ. Furthermore, the soft parts secrete a mucus to which grains of sand adhere, forming a "sandtube" of an extremely ephemeral nature, which sometimes extends over part of the valves.

GENERIC INDEX.

REFERENCE TO PLATES.

THE END

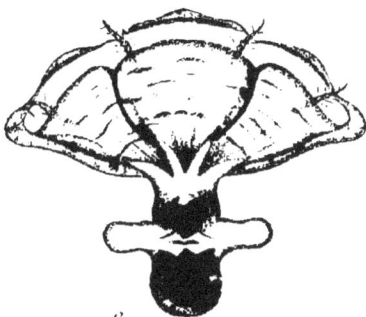

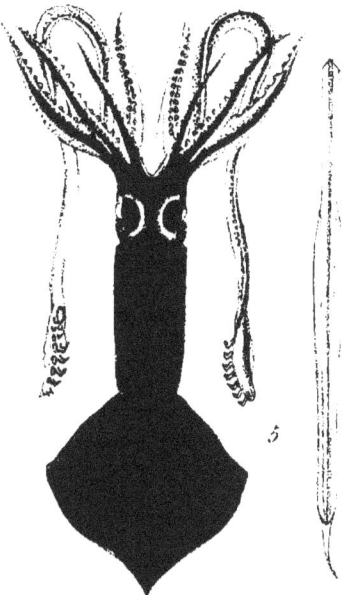

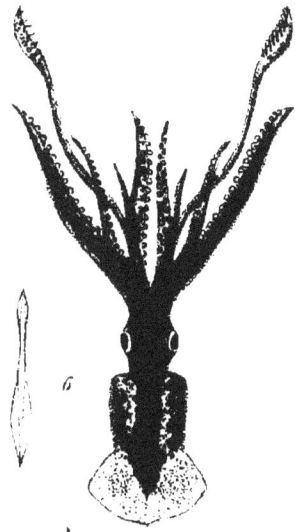

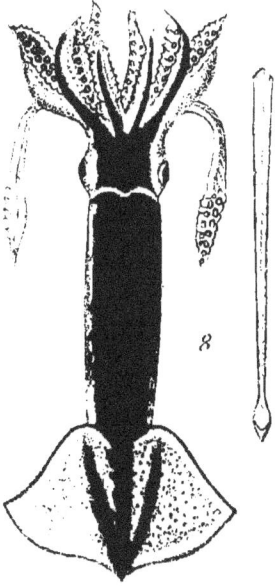

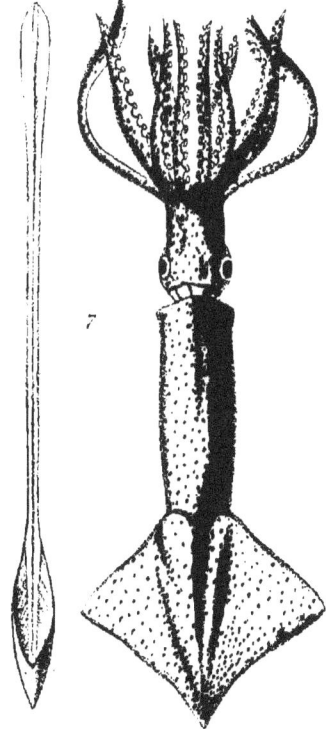

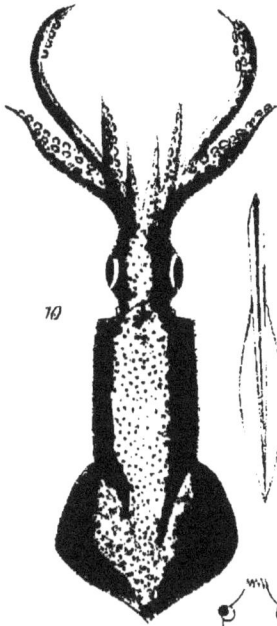
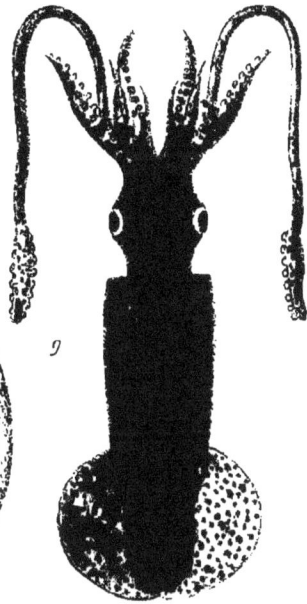

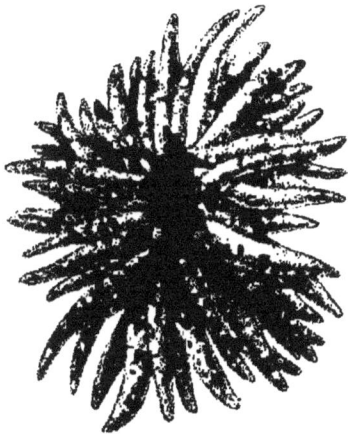

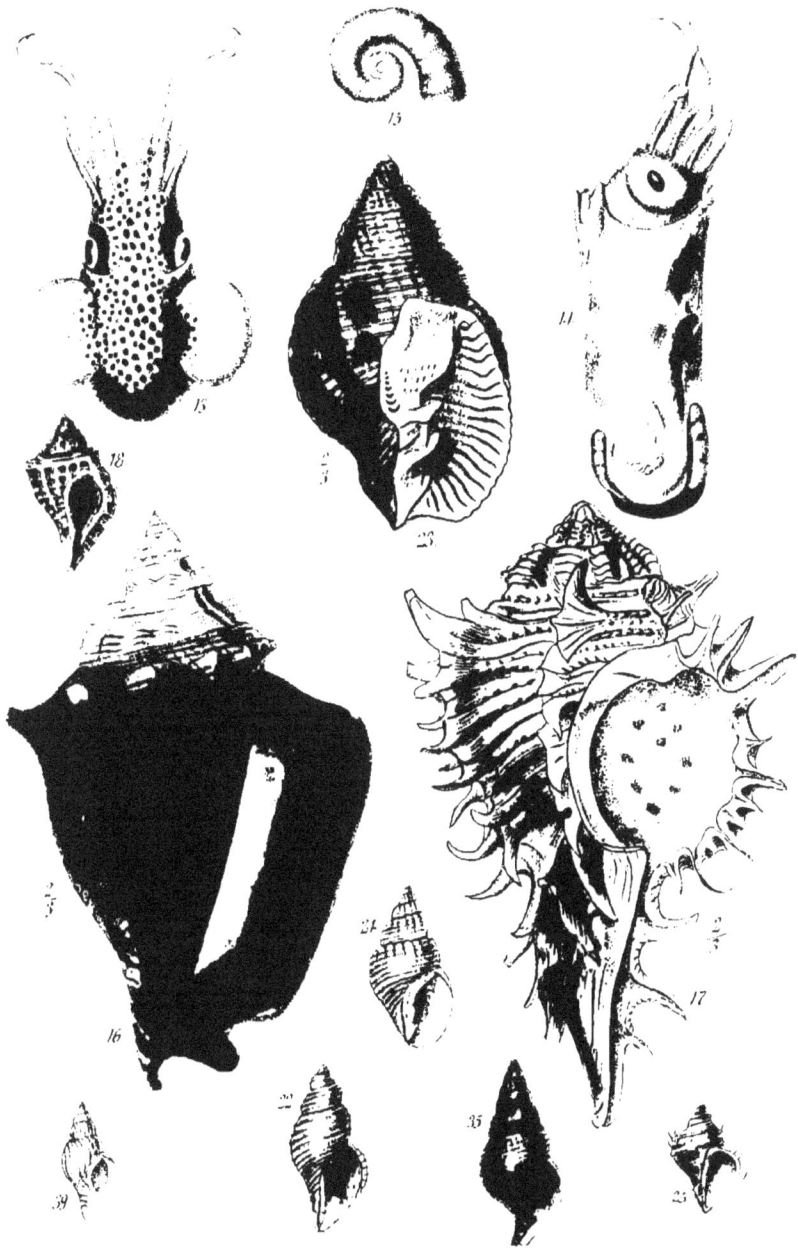

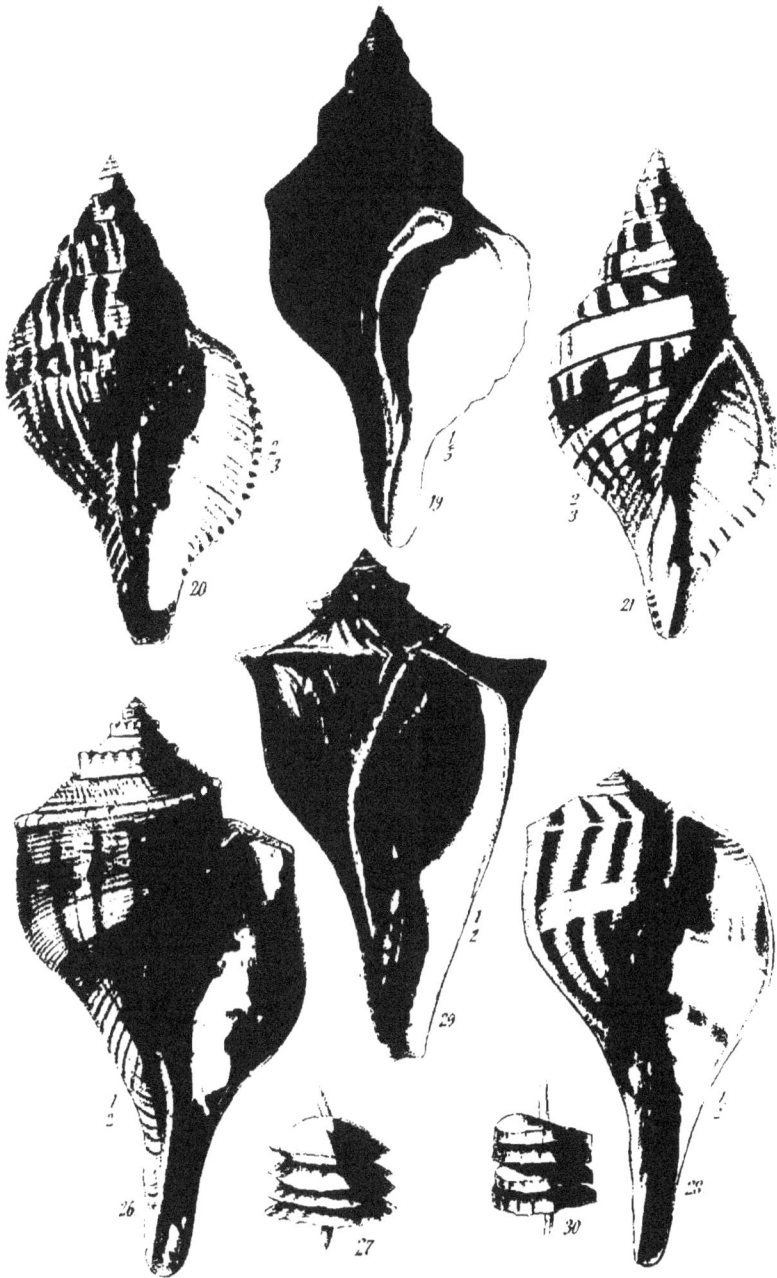

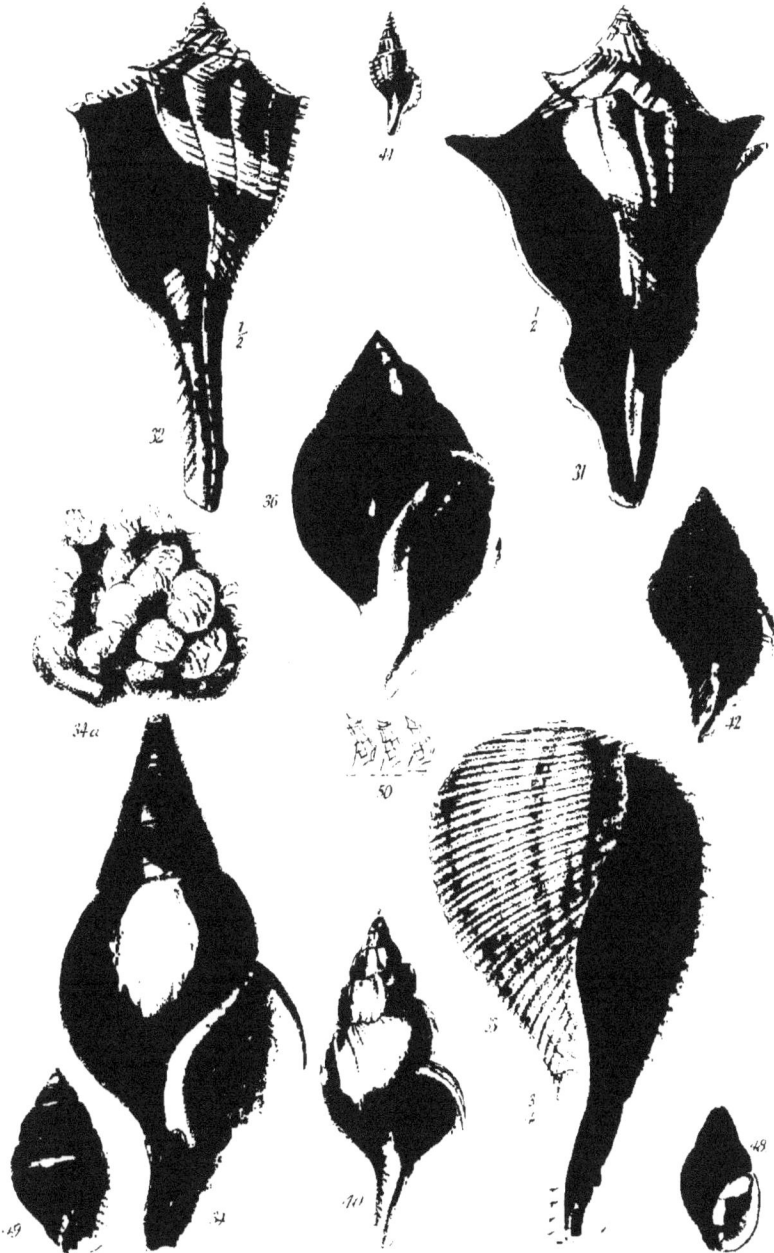

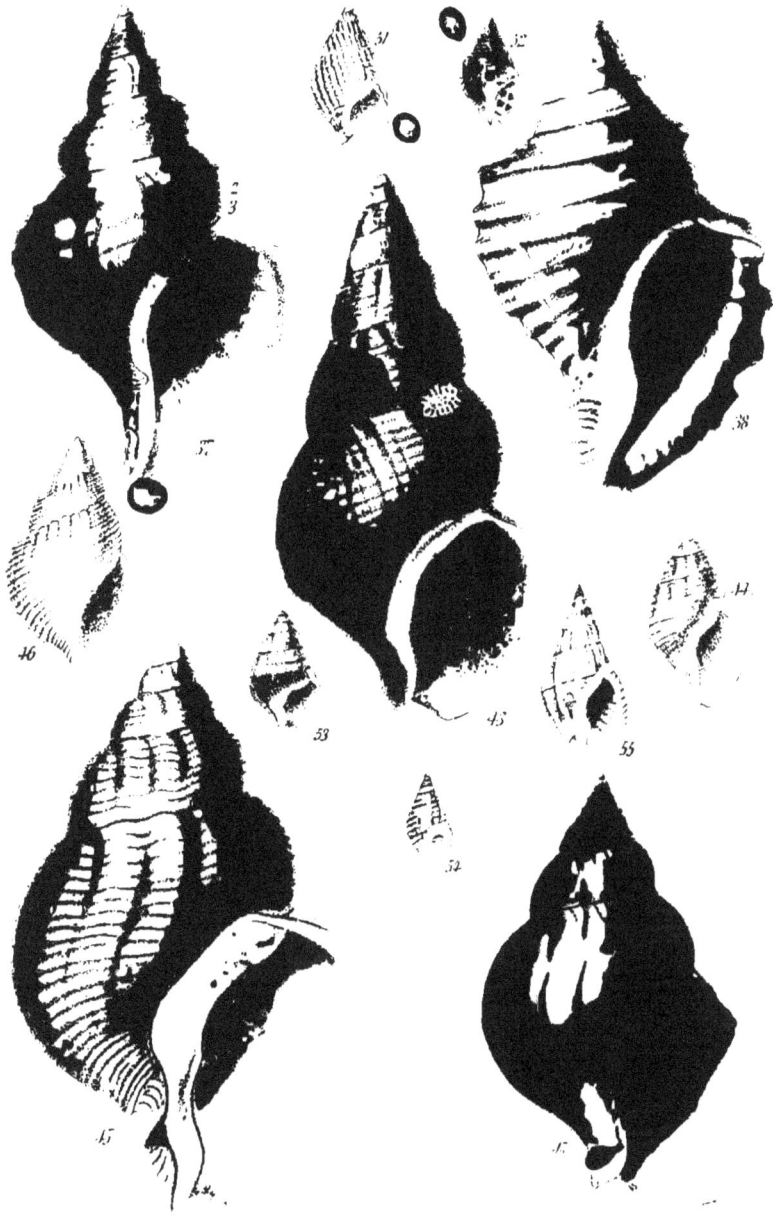

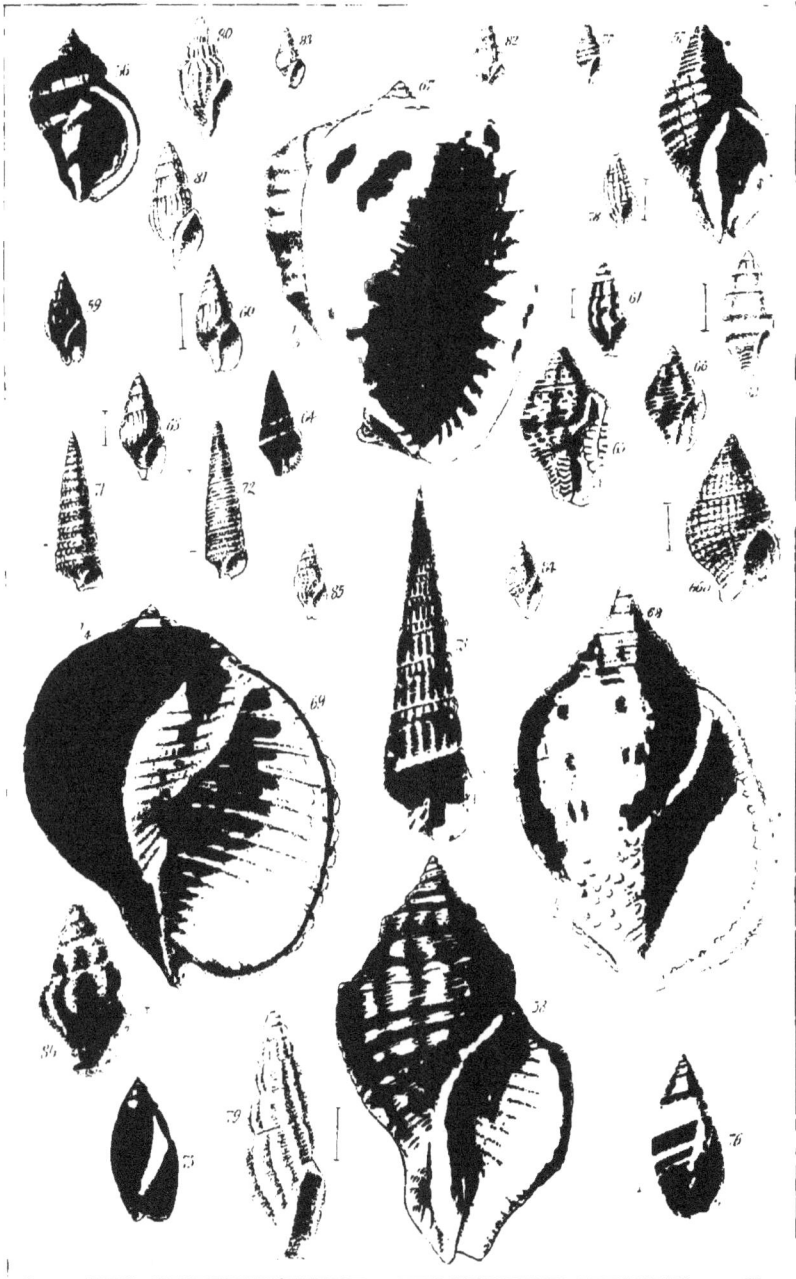

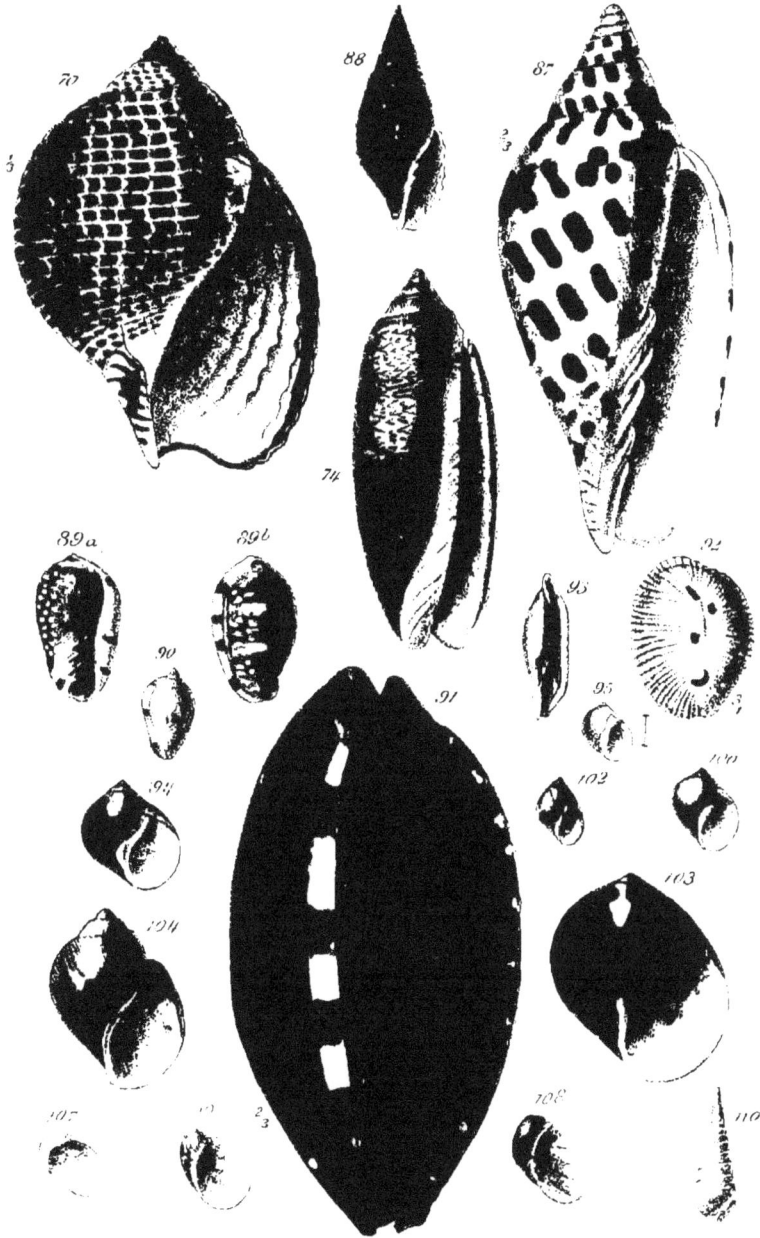

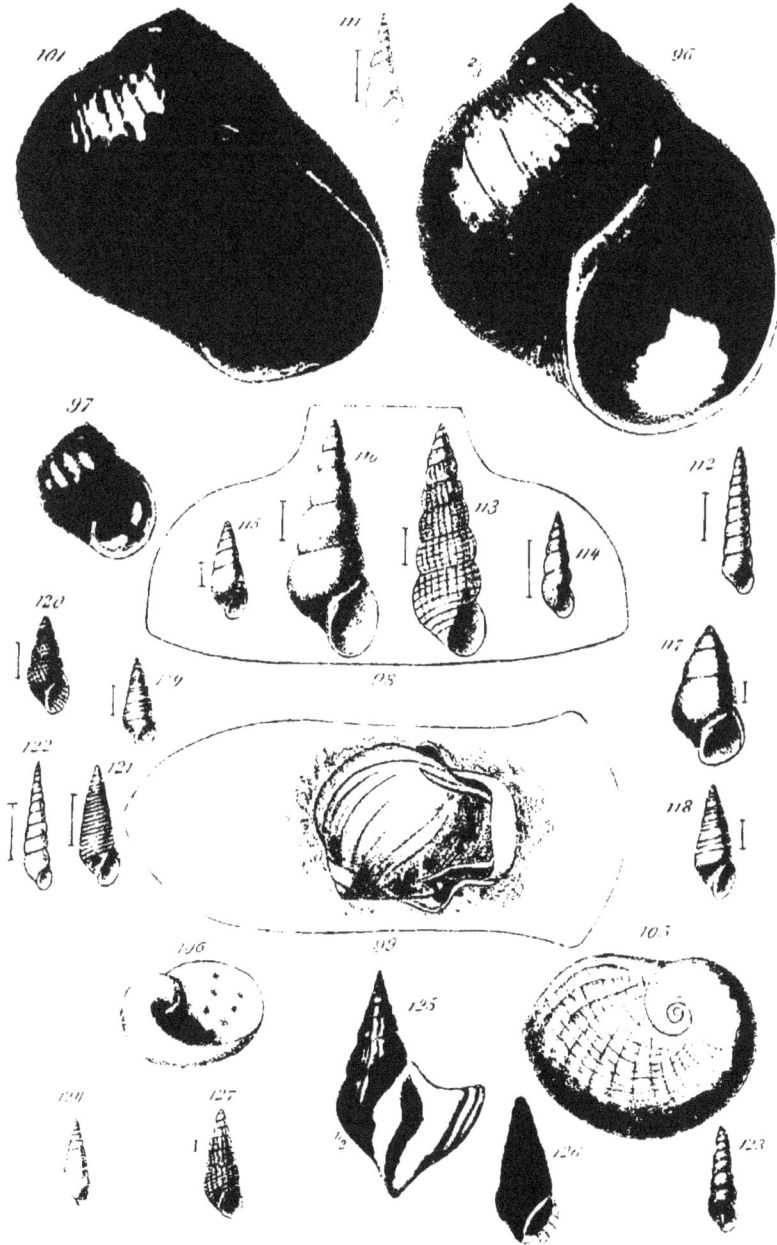

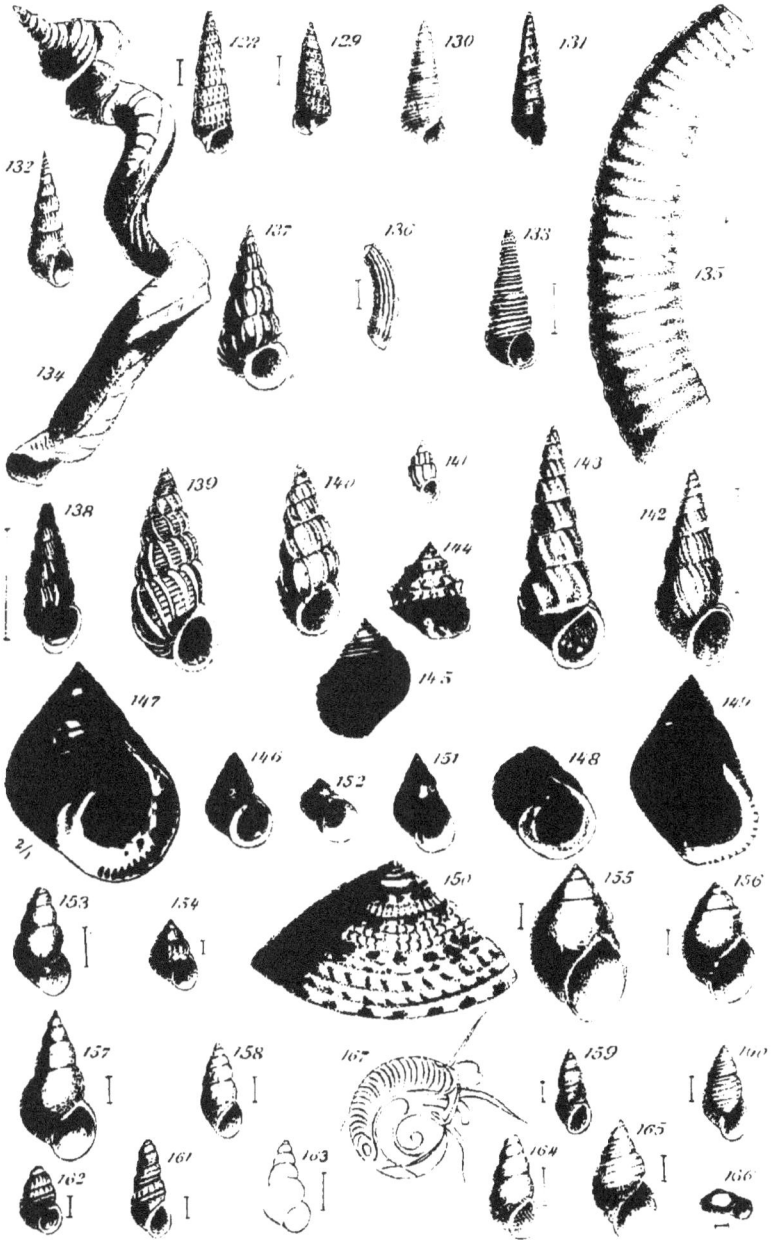

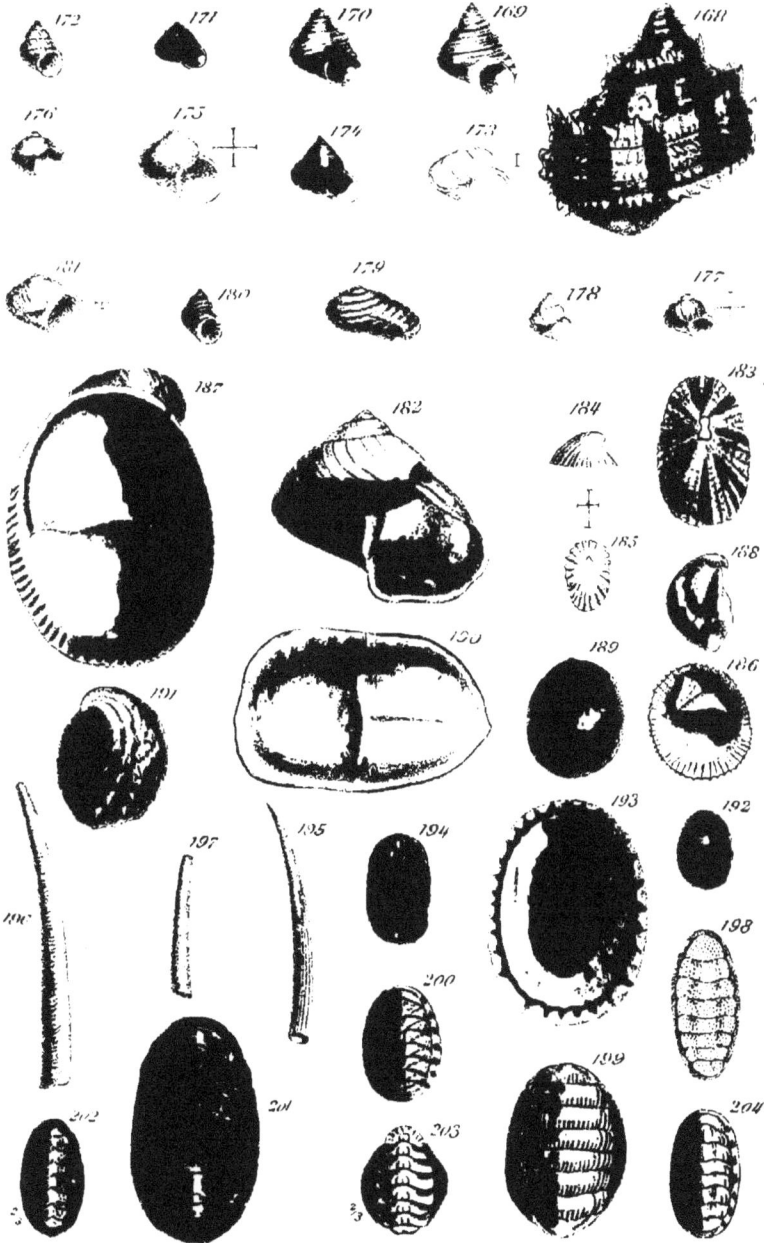

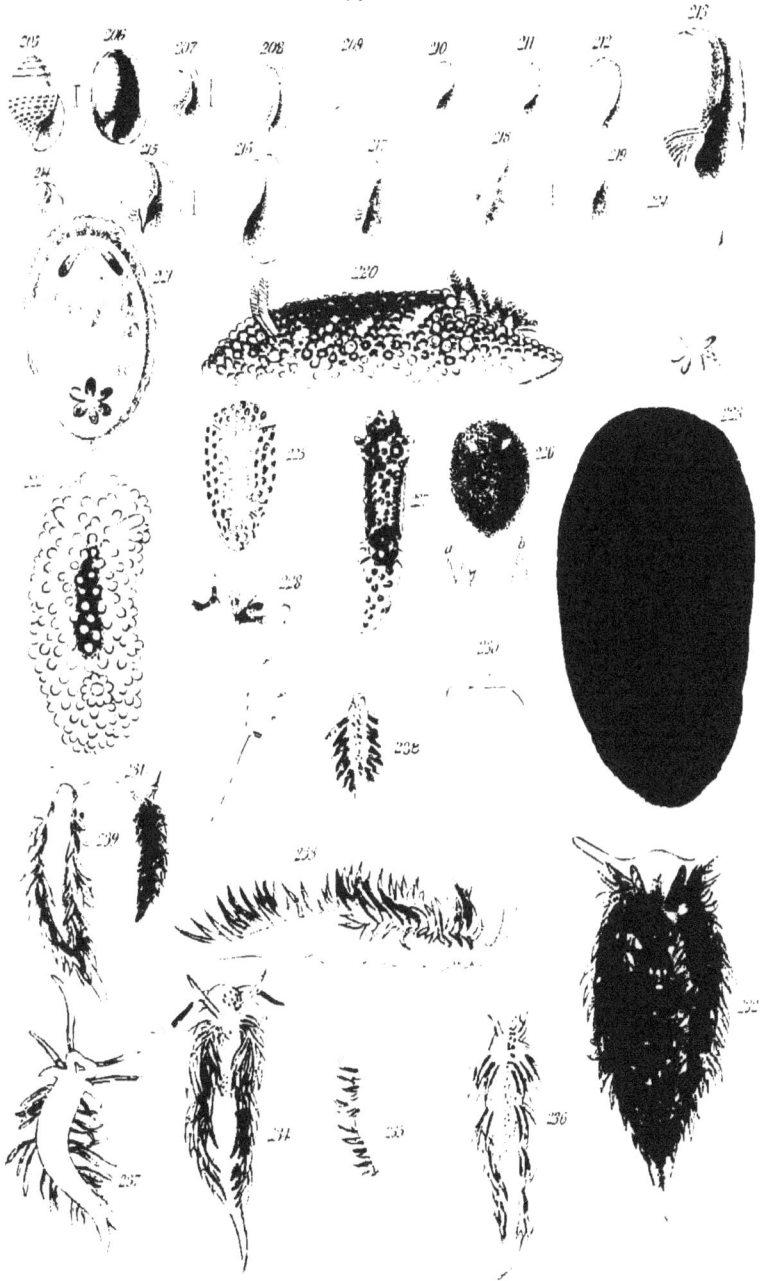

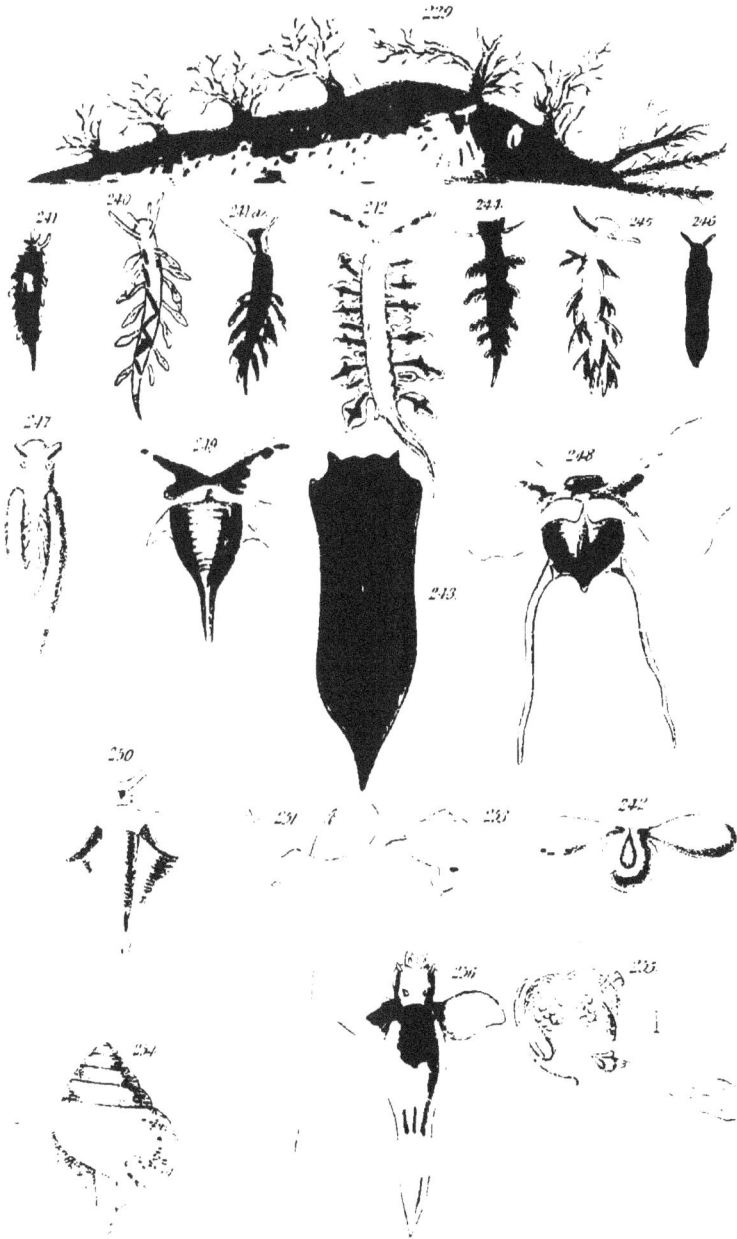

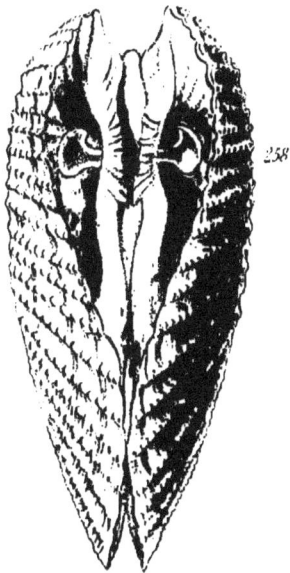

258

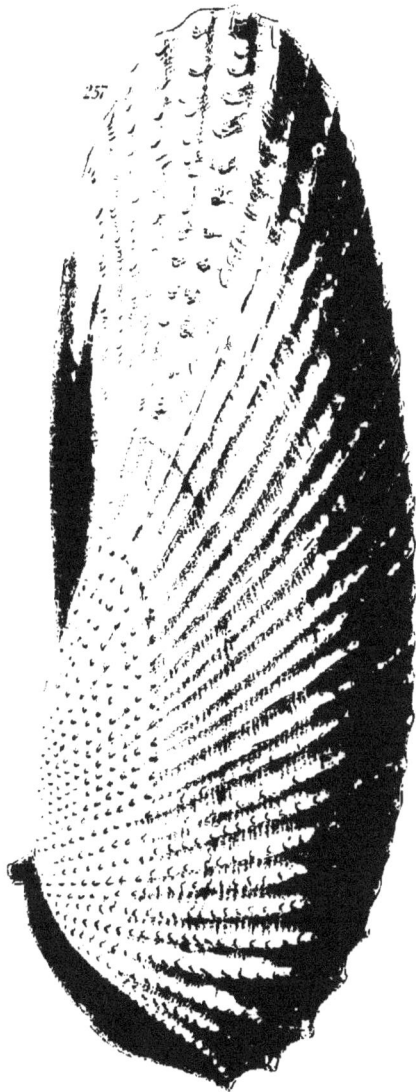

257

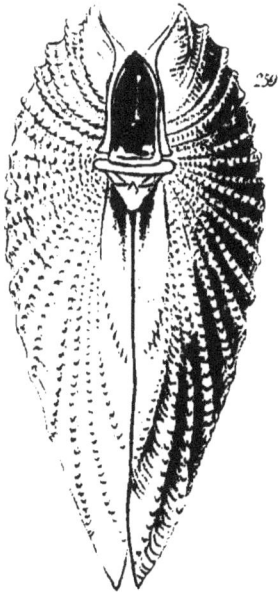

259

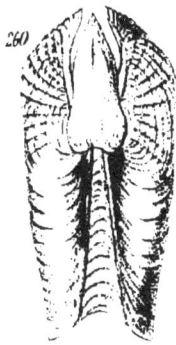

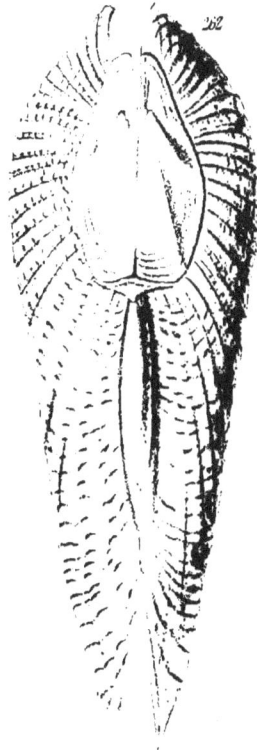

267

270

263  271  268

272 273 276

277

274

278  279

275

280

281 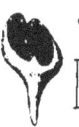 282

283

285 284 286

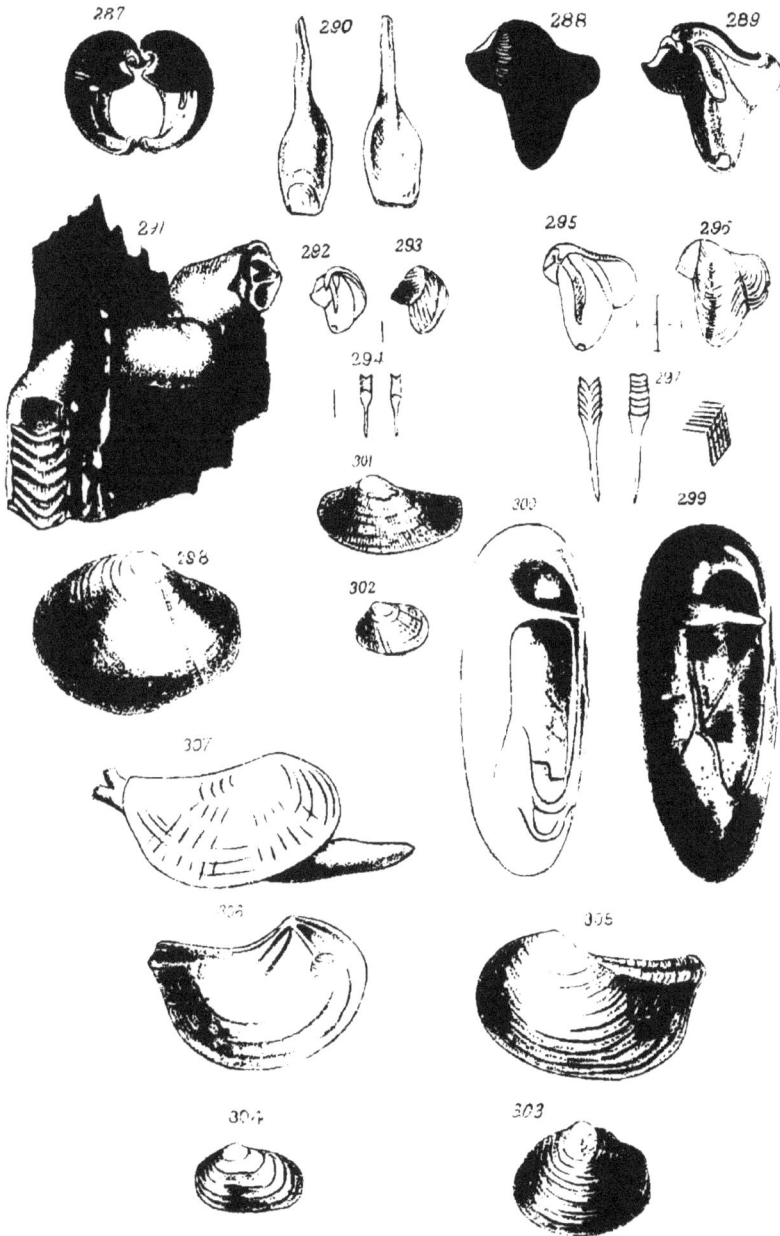

287

290

288

289

291

292
293

295

296

294

297

299

301

303

298

302

307

306

305

304

303

310

308

309

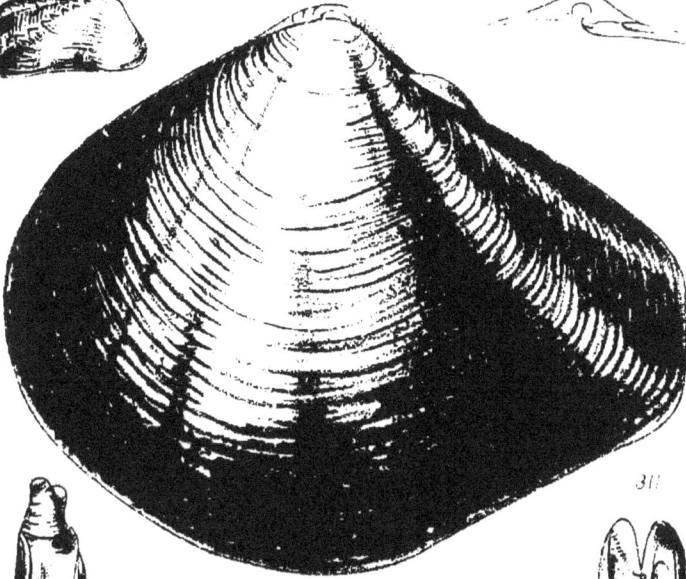

313

312

311

314

3/4

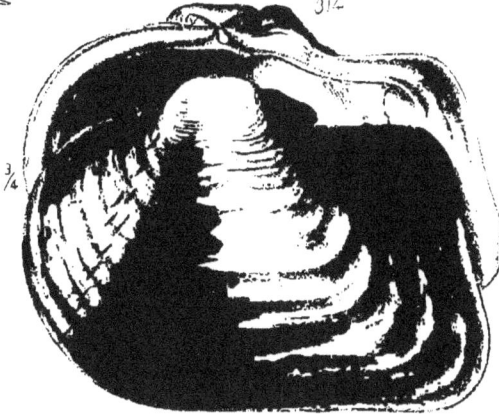

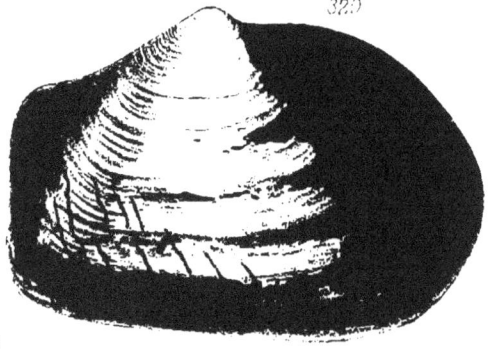

315

320

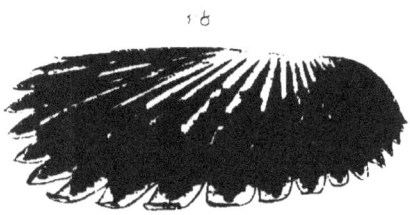

318

317

319

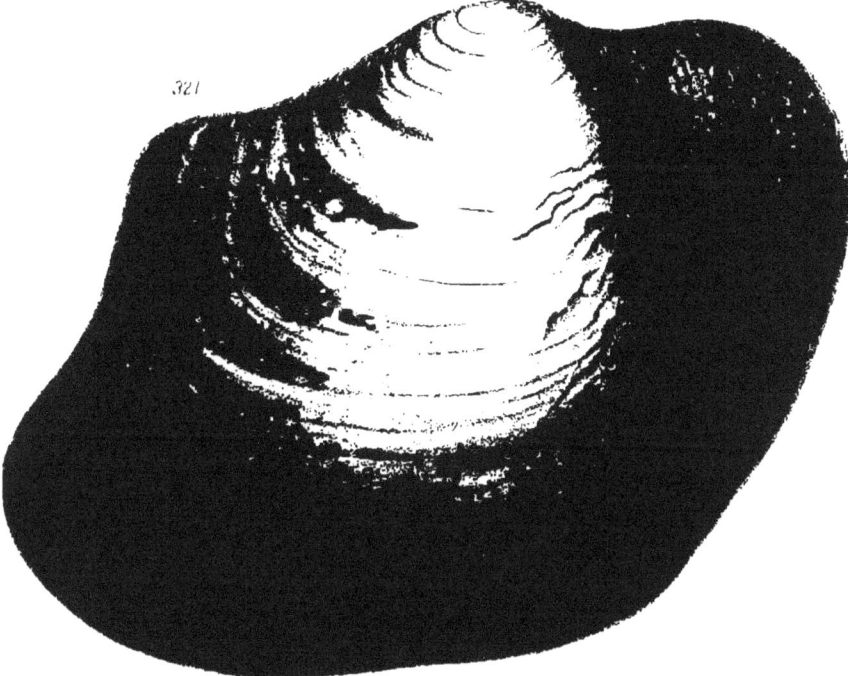

321

323

324

322

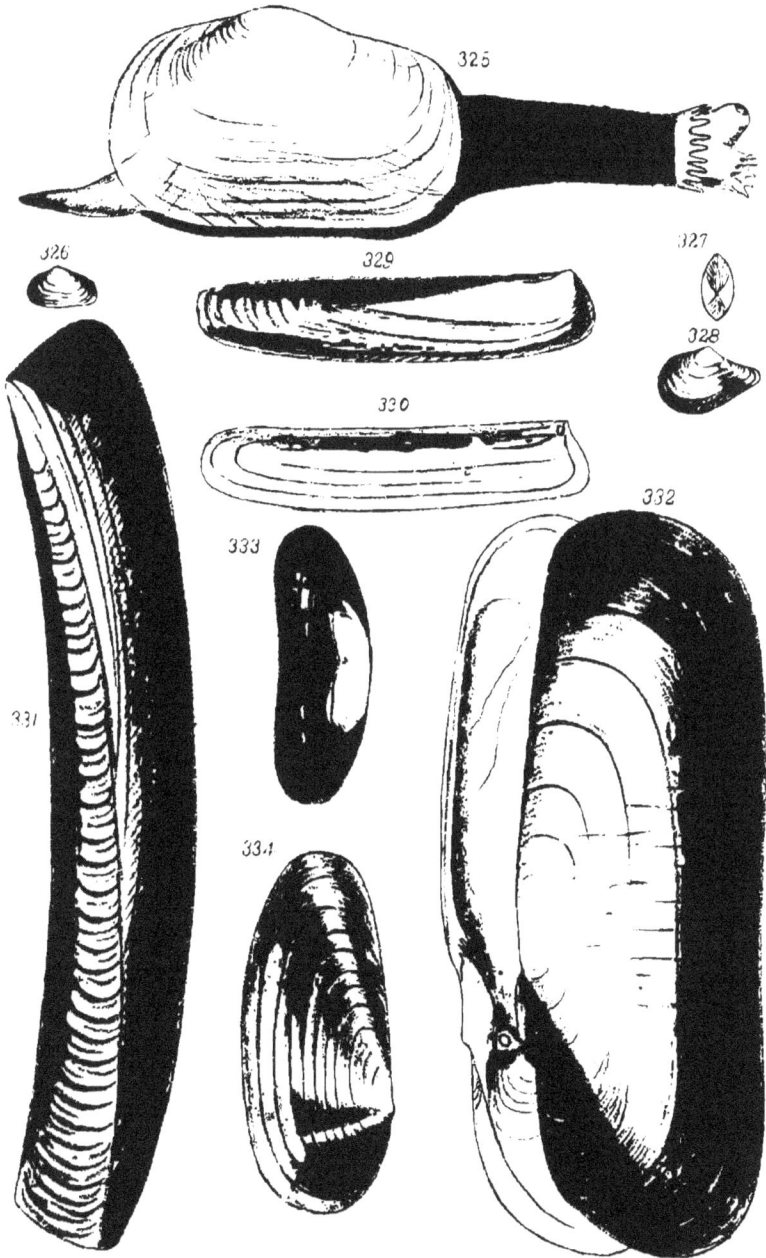

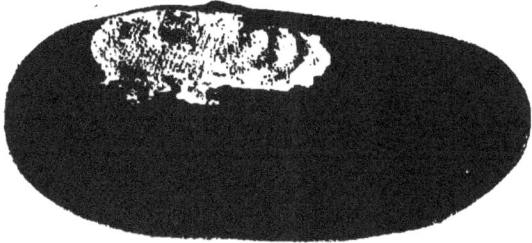

335

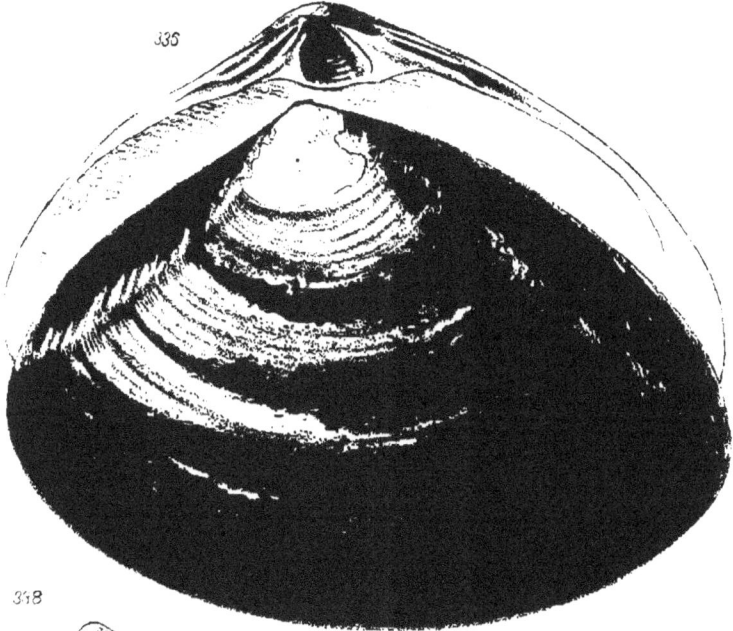

336

338

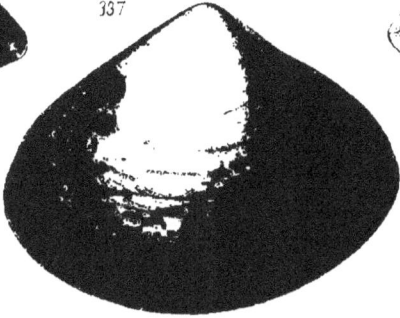

337

339

341

340

84.

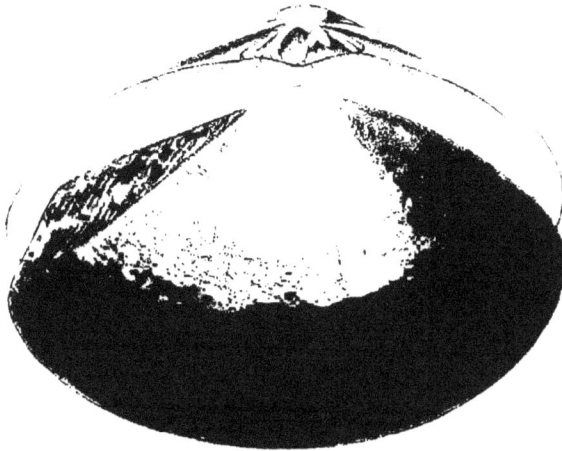

342

346

348

34~

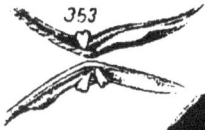

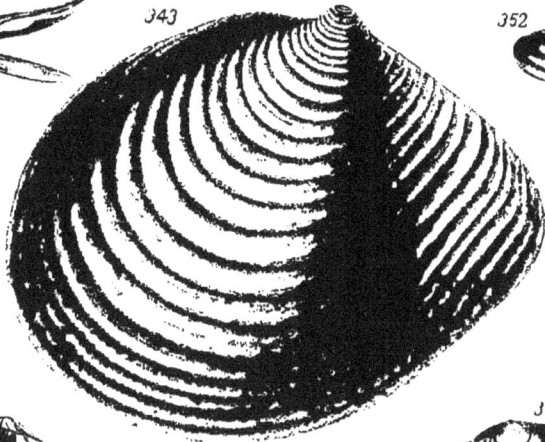

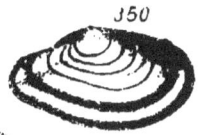

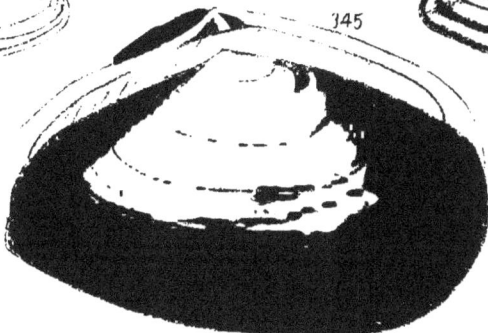

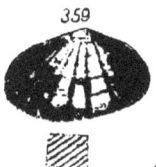

354

356

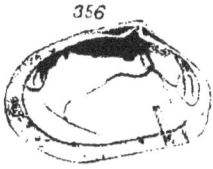

355

357

360

361

364

362

365

363

366

367

369

368

372

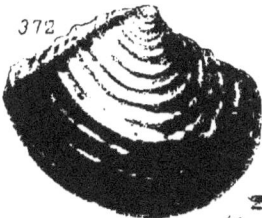

370

371

373

374

375

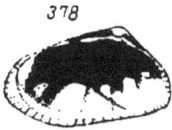

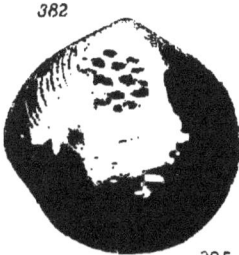

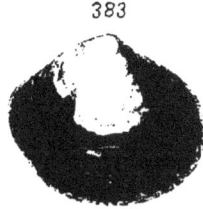

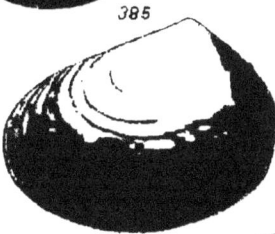

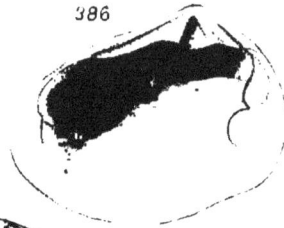

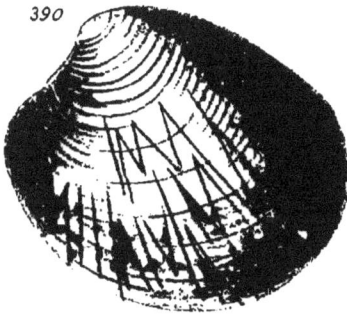

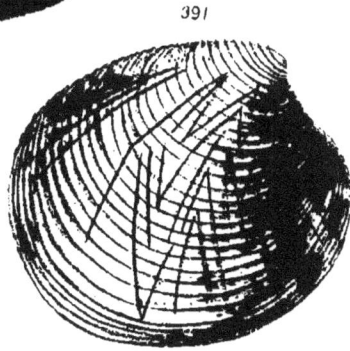

388

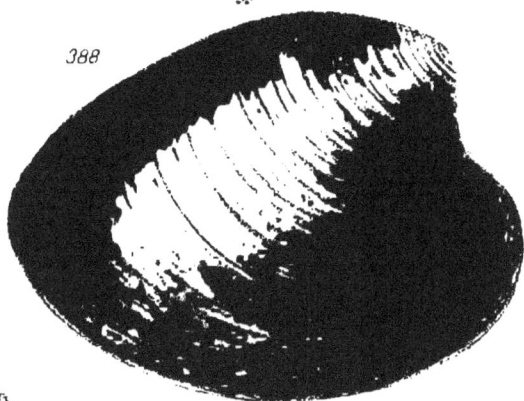

300

394

389

392

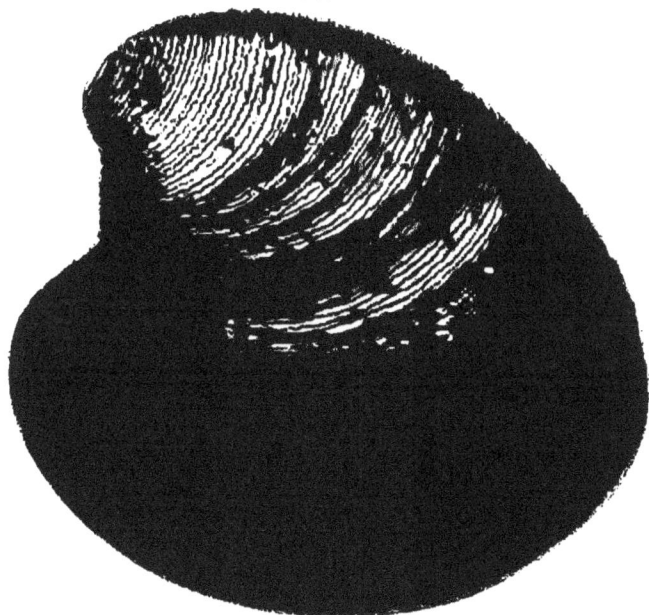

395

398

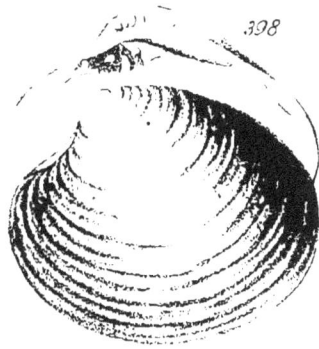

396

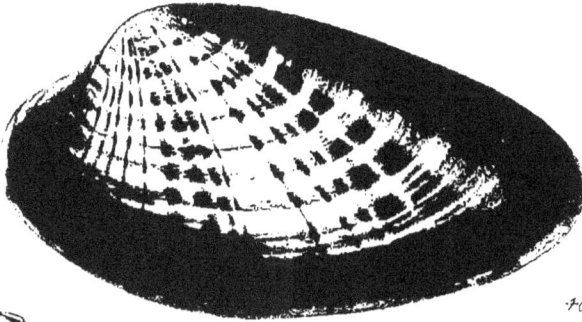

401

400

397

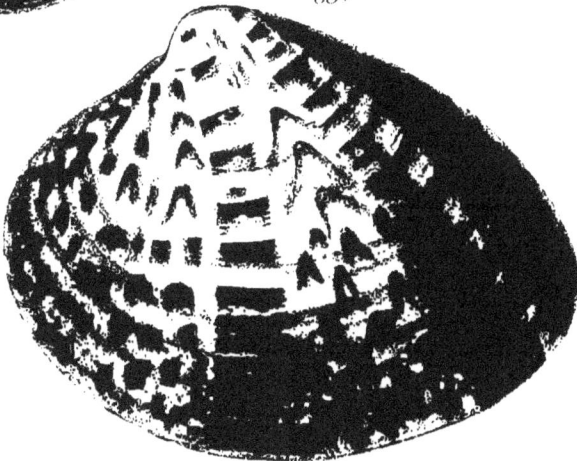

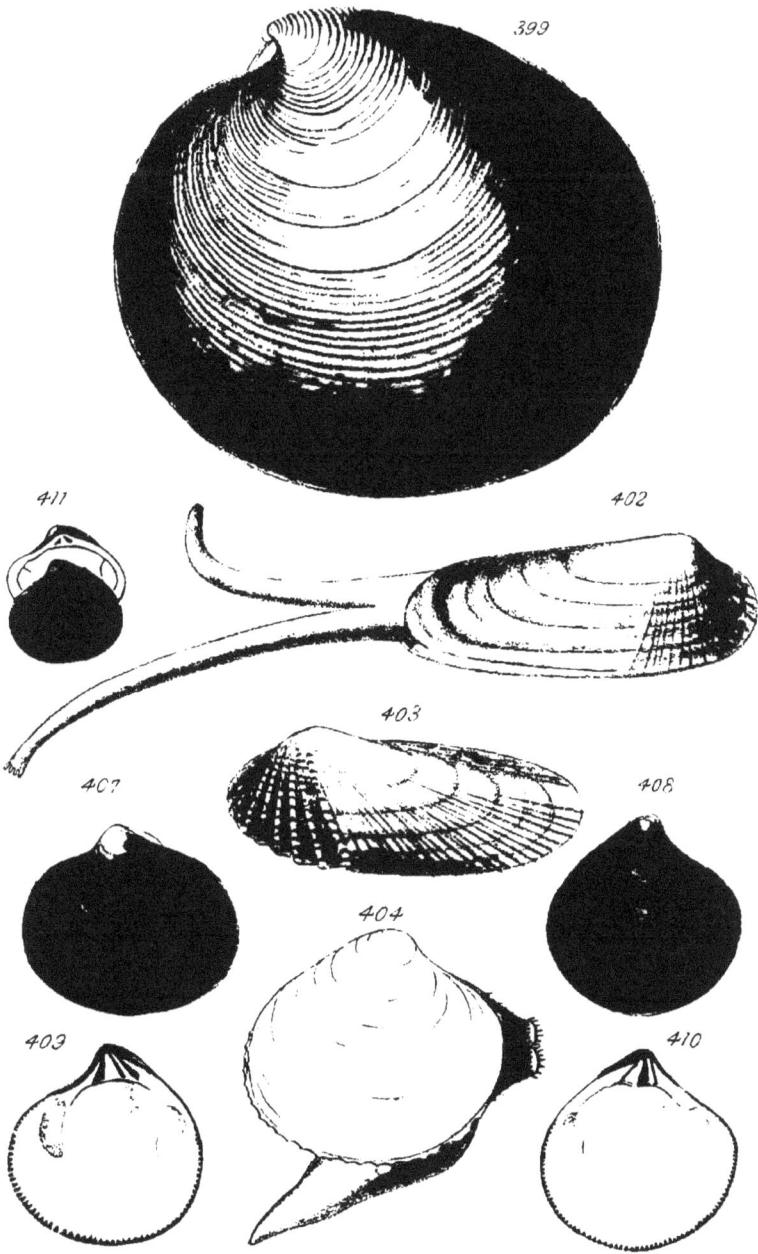

399

411 402

407 403 408

404

403 410

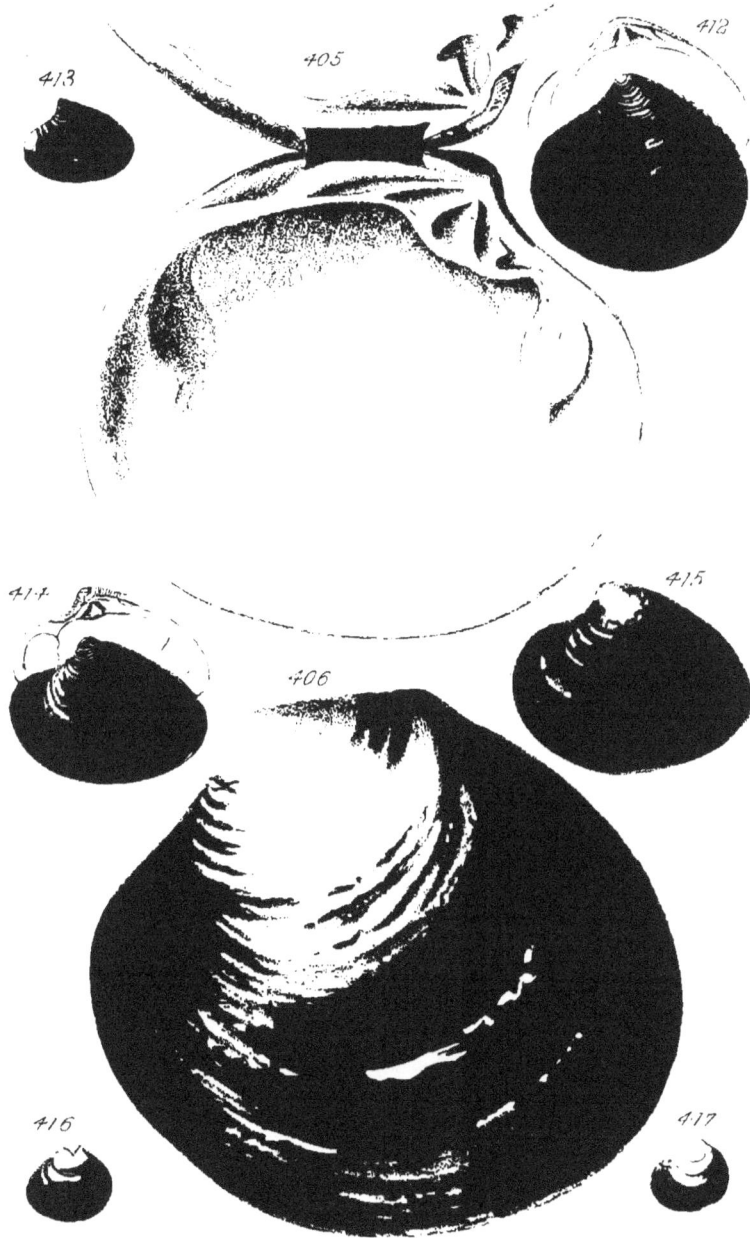

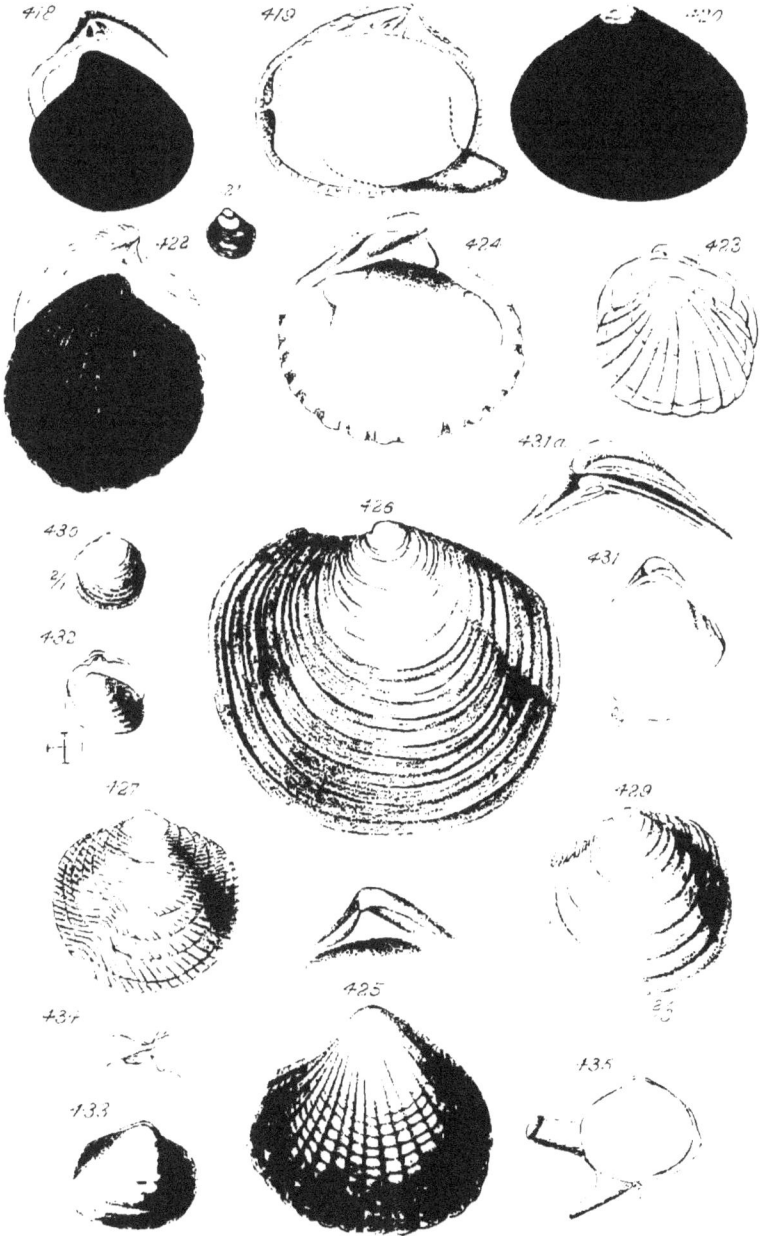

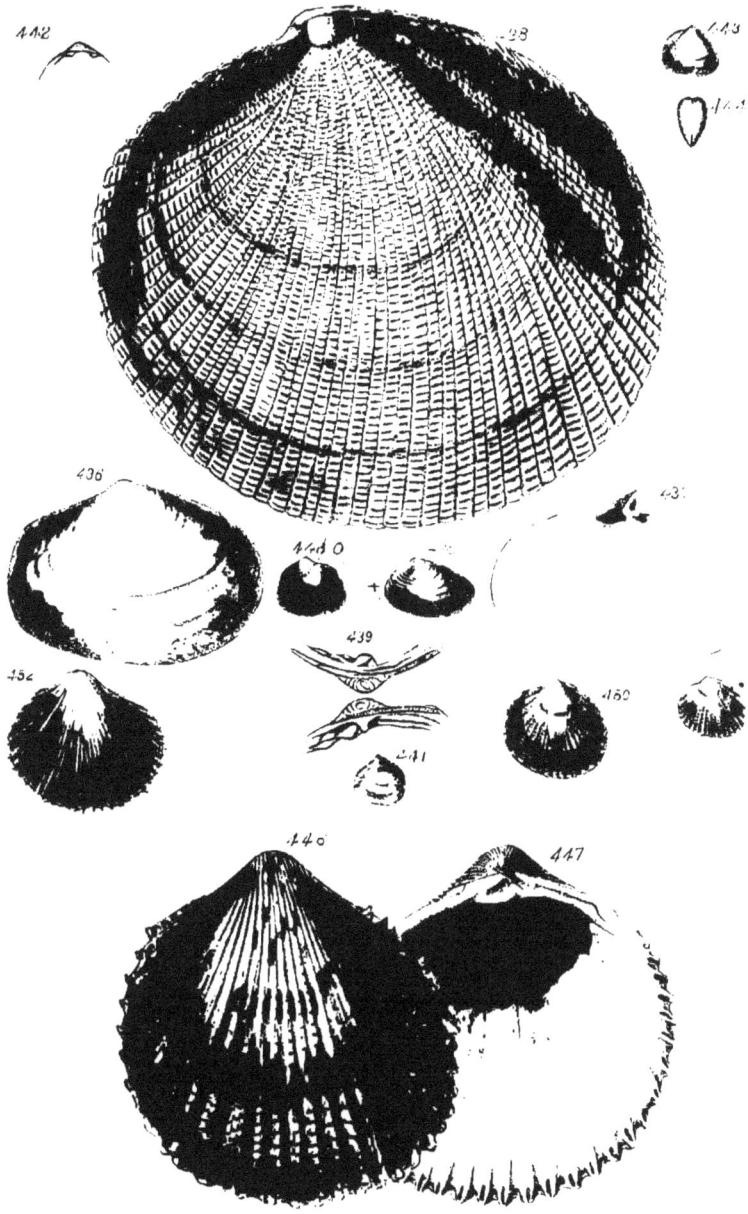

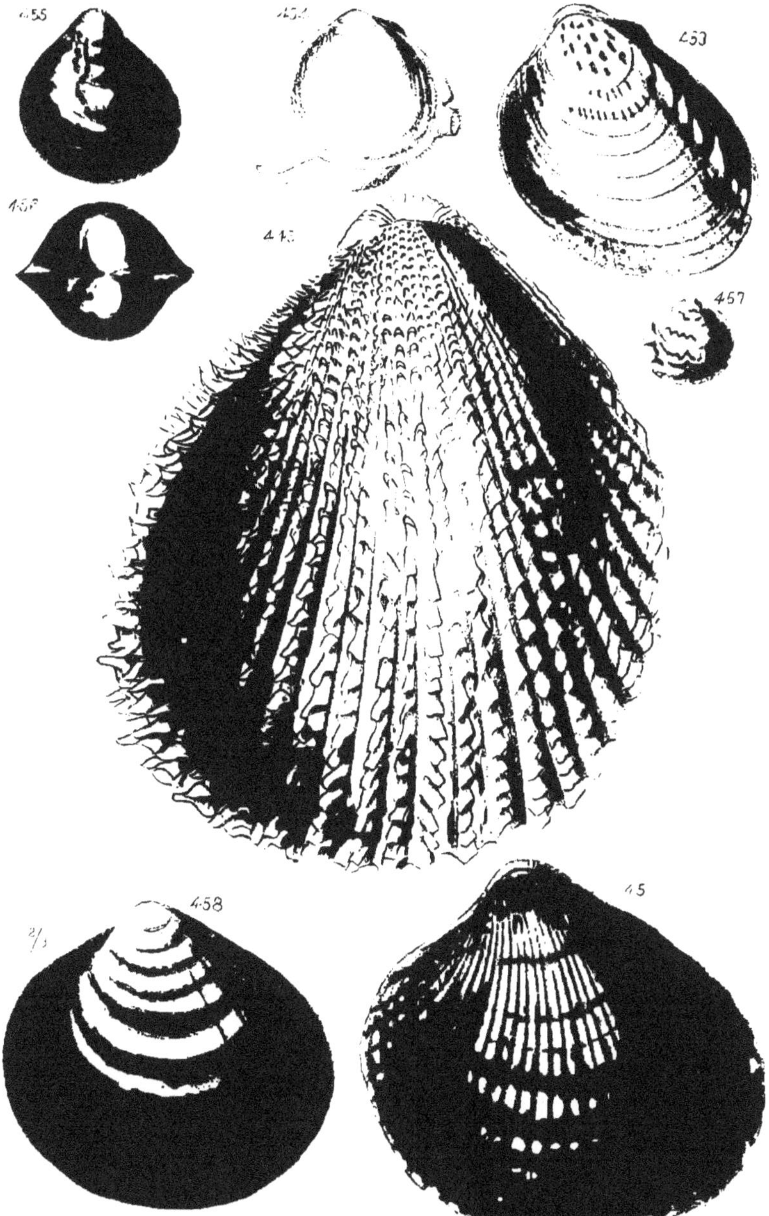
455
453
457
458
2/3

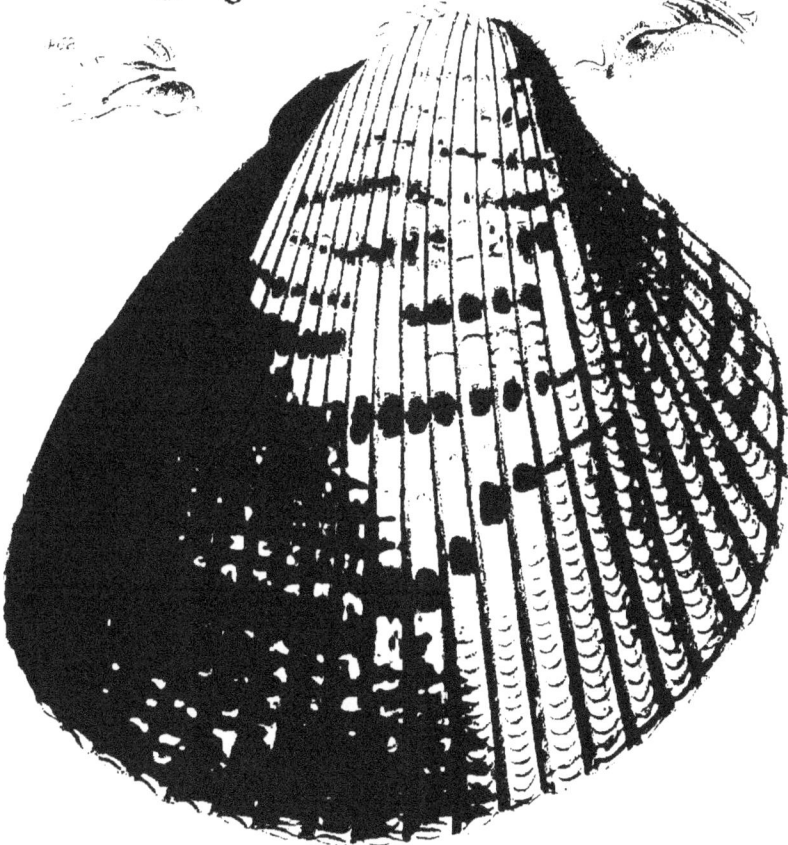

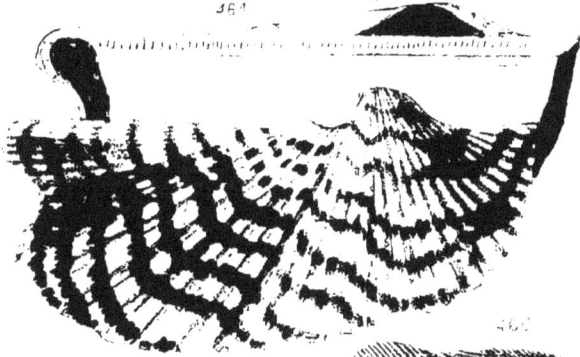

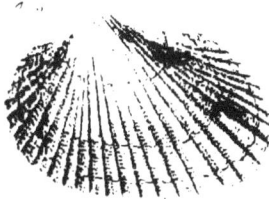

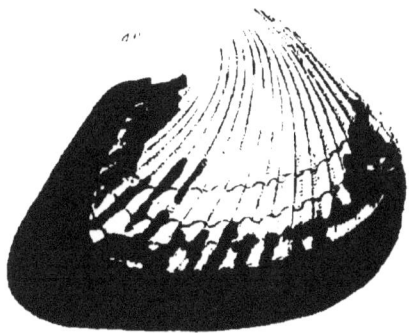
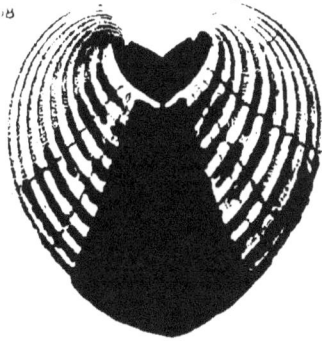

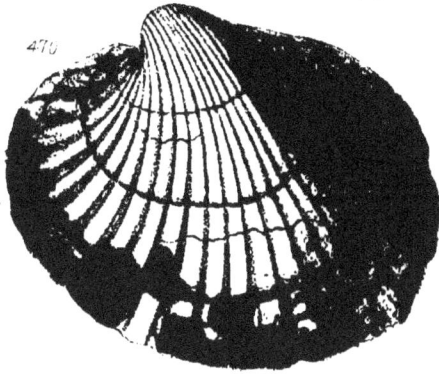

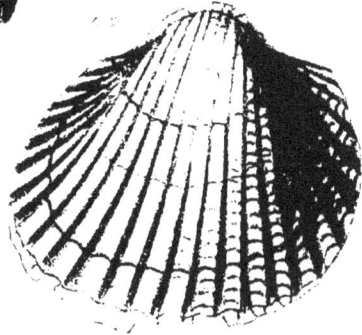

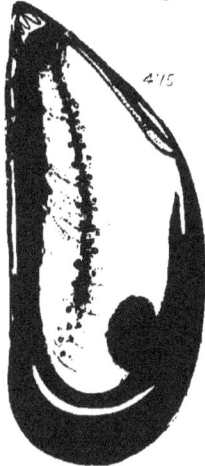

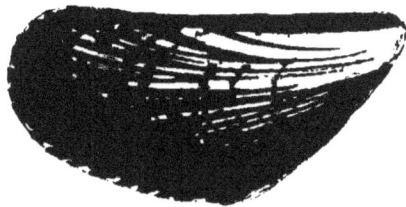

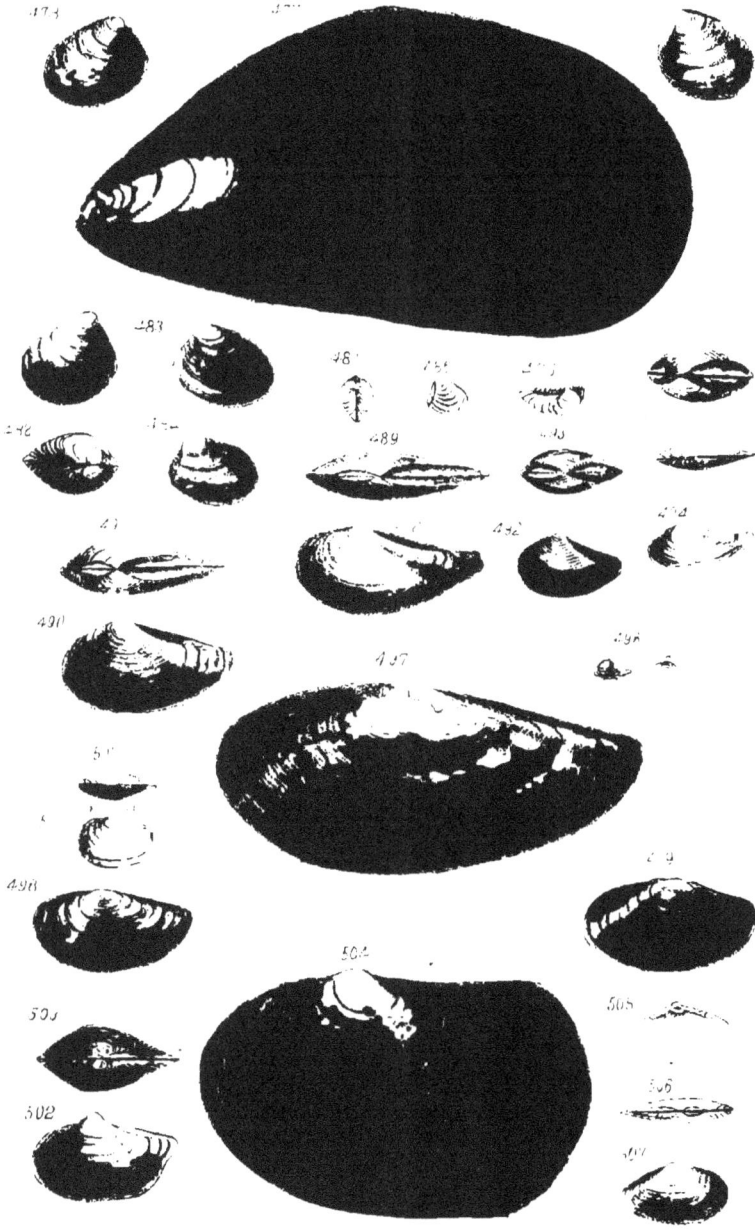

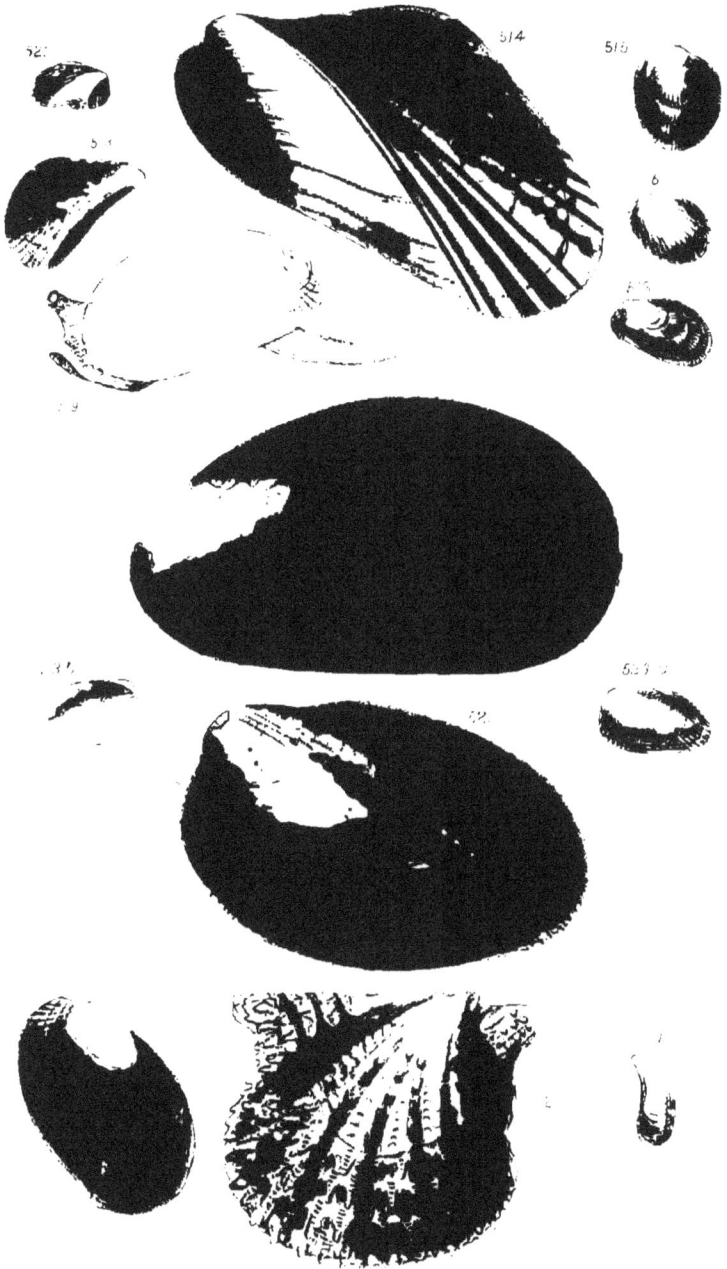

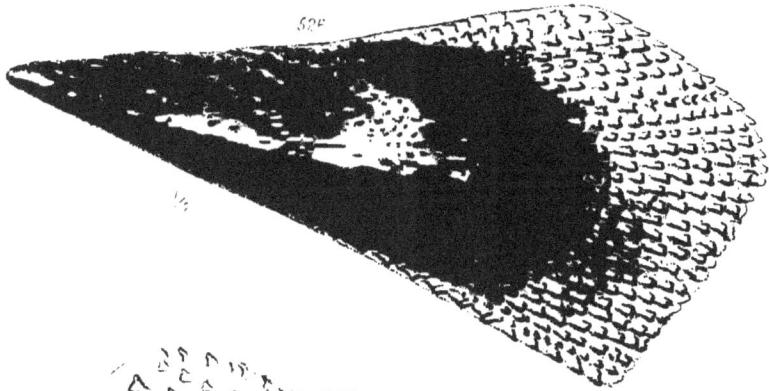

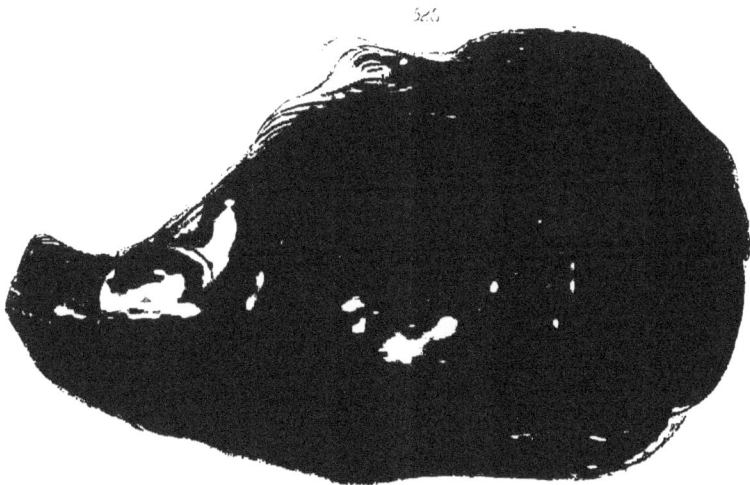

529

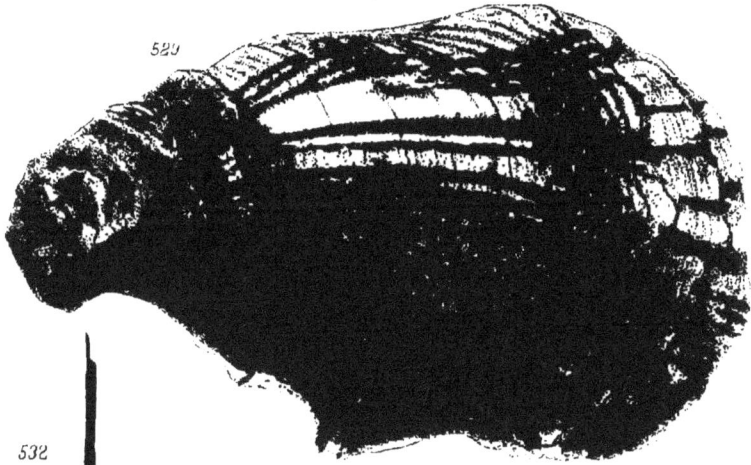

532

530

53

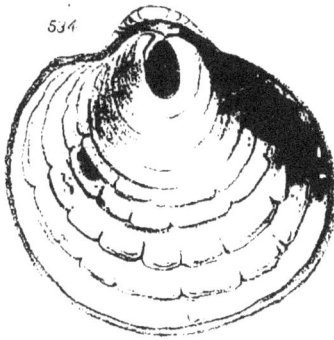

534

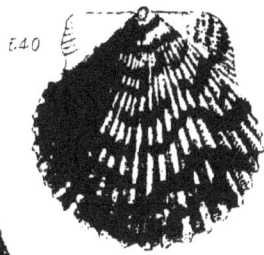

540

543

536

535

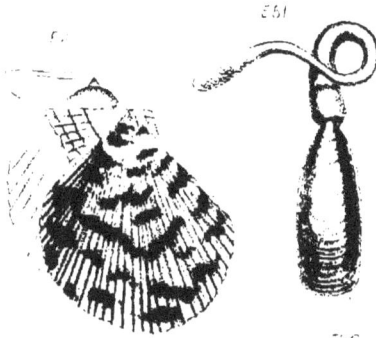

531

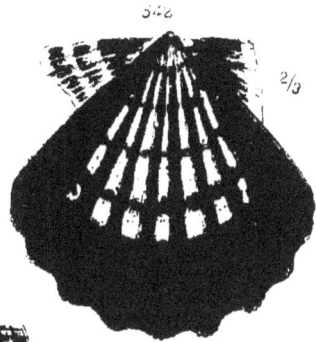

542

2/3

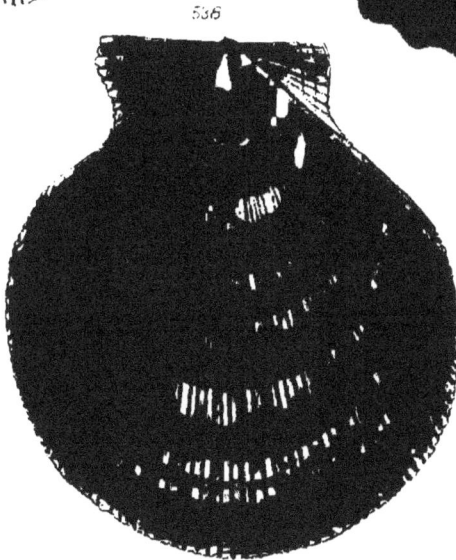

538

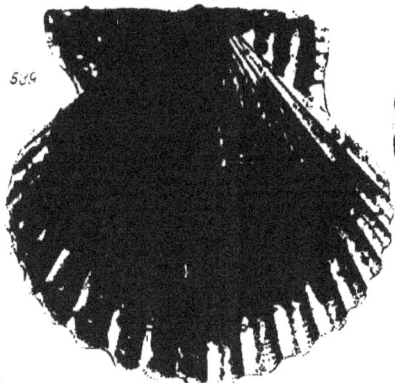

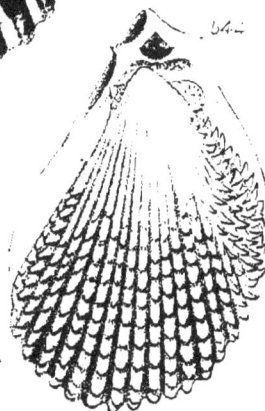

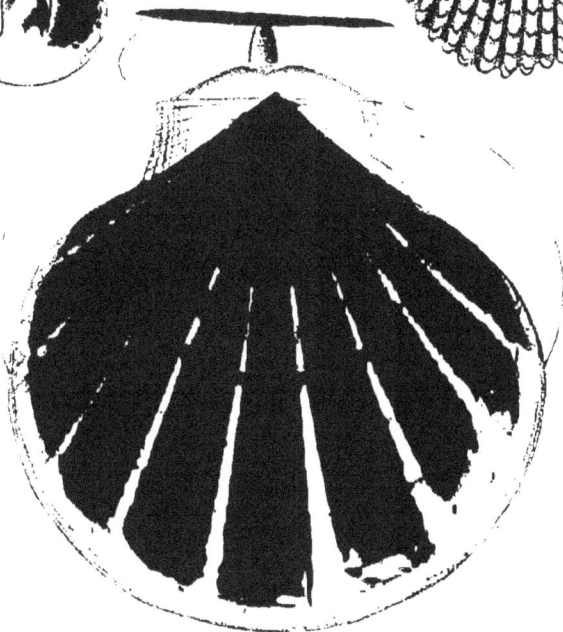

AMERICAN

MARINE CONCHOLOGY:

OR,

DESCRIPTIONS OF THE SHELLS

OF THE

ATLANTIC COAST OF THE UNITED STATES,

FROM MAINE TO FLORIDA.

BY

GEORGE W. TRYON, JR.,

MEMBER OF THE ACADEMY OF NATURAL SCIENCES OF PHILADELPHIA.

PHILADELPHIA:

PUBLISHED BY THE AUTHOR,

No. 19 North Sixth Street.

PROSPECTUS.

THE undersigned proposes to publish the AMERICAN MARINE CONCHOLOGY by subscription, provided that a sufficient number of names are obtained to guarantee to him a reasonable return for the expenditure incurred. The edition will be limited to *one hundred copies*, and the lithographic drawings will be destroyed when that number of impressions have been printed from them. Subscribers will thus obtain a work which cannot fail to become scarce and valuable in a short period.

The AMERICAN MARINE CONCHOLOGY will be published in octavo Parts, each containing 32 pages of text, beautifully printed on fine paper, and eight lithographic plates, crowded with figures; representing in all about six hundred species of shells, many of which will be described for the first time as inhabitants of our Atlantic coast.

CONDITIONS.

The FIRST PART will be ready about January 1st, 1873, and the work will be completed in five or six Parts, published quarterly.

1. Twenty-five copies will be prepared, having in addition to the finely colored plates, a duplicate set of plates printed on tinted paper. The text for this *edition de luxe* will be printed on very heavy cream-tinted paper.

2. Fifty copies will be printed on fine white paper, with the colored plates only.

3. Twenty-five copies will have uncolored illustrations.

PRICE PER PART, *Payable on Delivery:*

FINE EDITION, duplicate Plates, etc. .	Seven Dollars.
COLORED EDITION	Five Dollars.
UNCOLORED EDITION	Three Dollars.

Gentlemen desiring to subscribe for the work will please sign the accompanying obligation and return it immediately to

GEO. W. TRYON, JR.,
Academy of Natural Sciences, Philadelphia.

www.ingramcontent.com/pod-product-compliance
Lightning Source LLC
Chambersburg PA
CBHW021508210326
41599CB00012B/1184